SpringerBriefs in Environmental Science

SpringerBriefs in Environmental Science present concise summaries of cutting-edge research and practical applications across a wide spectrum of environmental fields, with fast turnaround time to publication. Featuring compact volumes of 50 to 125 pages, the series covers a range of content from professional to academic. Monographs of new material are considered for the SpringerBriefs in Environmental Science series.

Typical topics might include: a timely report of state-of-the-art analytical techniques, a bridge between new research results, as published in journal articles and a contextual literature review, a snapshot of a hot or emerging topic, an in-depth case study or technical example, a presentation of core concepts that students must understand in order to make independent contributions, best practices or protocols to be followed, a series of short case studies/debates highlighting a specific angle.

SpringerBriefs in Environmental Science allow authors to present their ideas and readers to absorb them with minimal time investment. Both solicited and unsolicited manuscripts are considered for publication.

More information about this series at http://www.springer.com/series/8868

Davide Geneletti • Chiara Cortinovis
Linda Zardo • Blal Adem Esmail

Planning for Ecosystem Services in Cities

Davide Geneletti
University of Trento
TRENTO, Italy

Chiara Cortinovis
University of Trento
TRENTO, Italy

Linda Zardo
University of Trento
TRENTO, Italy

Blal Adem Esmail
University of Trento
TRENTO, Italy

This book is an open access publication.

ISSN 2191-5547   ISSN 2191-5555   (electronic)
SpringerBriefs in Environmental Science
ISBN 978-3-030-20023-7   ISBN 978-3-030-20024-4   (eBook)
https://doi.org/10.1007/978-3-030-20024-4

This Springer imprint is published by the registered company Springer Nature Switzerland AG
The registered company address is: Gewerbestrasse 11, 6330 Cham, Switzerland

Acknowledgments

Research for this book has been partly conducted for the projects ESMERALDA and ReNature, receiving funding from the European Union's Horizon 2020 research and innovation programme under grant agreements No 642007 and No 809988, respectively. We also acknowledge support from the Italian Ministry of Education, University and Research (MIUR) in the frame of the "Departments of Excellence" grant L. 232/2016.

Contents

About the Authors

Davide Geneletti, PhD associate professor at the University of Trento, leader of the Planning for Ecosystem Services research group (www.planningfores.com), former research fellow at Harvard University's Sustainability Science Program, and visiting scholar at Stanford University

Chiara Cortinovis, PhD postdoc fellow at the Department of Civil, Environmental and Mechanical Engineering of the University of Trento, former visiting scholar at Humboldt University of Berlin, expert for the European Commission, and planning consultant for Public Administrations

Linda Zardo, PhD environmental expert for the United Nations Human Settlements Programme in Southeast Africa and former research fellow at the Department of Civil, Environmental and Mechanical Engineering of the University of Trento

Blal Adem Esmail, PhD postdoc fellow at the Department of Civil, Environmental and Mechanical Engineering of the University of Trento, former visiting scholar at Leibniz University Hannover, teaching assistant in Water Engineering, and consultant in the fields of Building Construction and Oil & Gas Research

Chapter 1
Introduction

1.1 Ecosystem Services in Decision-Making

Human life on Earth depends on ecosystems. This is the main message conveyed by the concept of ecosystem services (ES), which has gained an ever-increasing attention in the scientific (McDonough et al. 2017) and policy debate (e.g., CBD 2011; European Commission 2006, 2010) of the last two decades. The success of the term 'ecosystem services' is arguably due to its encompassing all "the direct and indirect contributions of ecosystems to human wellbeing" (TEEB 2010a), thus providing a comprehensive framework to describe the multiple relationships between humans and nature.

The term 'ecosystem services' appeared for the first time in 1981 in a book by Ehrlich and Ehrlich as an evolution of the term 'environmental services' (Ehrlich and Ehrlich 1981), but it remained for some time confined within the disciplinary boundaries of conservation ecology. Only in the late nineties two pioneering works brought ES to the forefront of the scientific debate. In 1997, a comprehensive overview of the ES through which nature underpins human wellbeing was provided (Daily 1997), while a group of ecologists and economists made the first attempt to estimate the total economic value of the biosphere based on ES (Costanza et al. 1997), generating a rapidly-growing interest in the topic. In 2005, the publication of the *Millennium Ecosystem Assessment* report (MA 2005) under the umbrella of the United Nations Environmental Programme (UNEP) put ES high on the world policy agenda. The ES concept was proposed as an innovative way to communicate the growing concerns for the unprecedented rates of ecosystem degradation and biodiversity loss, thus providing an additional justification for nature conservation based on what nature does *for* people (Mace 2014, 2016).

What characterized the ES concept since its origin was the explicit link with decision-making. Gretchen Daily and colleagues identified in this link the main innovation of the ES approach, where ES values are acknowledged and assessed

© The Author(s) 2020
D. Geneletti et al., *Planning for Ecosystem Services in Cities*, SpringerBriefs
in Environmental Science, https://doi.org/10.1007/978-3-030-20024-4_1

with the specific purpose of informing decisions (Daily et al. 2009). Highlighting the dependency of human wellbeing on nature, the ES concept definitely makes clear that no trade-off should exist between sustainable human development and nature conservation (de Groot et al. 2010). Consequently, identifying, mapping, quantifying, and valuing ES is expected to improve decision making, ultimately promoting more sustainable development trajectories (TEEB 2010b; Díaz et al. 2015; Guerry et al. 2015). In the last years, efforts have been made to include ES in different decision-making processes to support the identification and comparison of costs and benefits of different policies (TEEB 2010b) and to contribute to the assessment of their impacts (Geneletti 2013).

At the international level, the acknowledgement of the need to secure a sustainable and fair provision of ES was explicitly at the basis of the adoption of the *Aichitargets* by the *Convention on Biological Diversity* (2010) and of the creation of the *Intergovernmental Science-Policy Platform on Biodiversity and Ecosystem Services* (2012). The European Union is at the forefront in pursuing these obligations and is leading the way toward mainstreaming the ES approach by progressively embedding the ES concept in its policies (Bouwma et al. 2017). Through the *EU Biodiversity strategy to 2020,* EU Member States committed to map and assess ES in their territory, thus setting the base for continuous monitoring and the inclusion of ES in the system of national accounting and reporting across the EU (Maes et al. 2012, 2016). Comprehensive ES assessments have also been carried out at national level, both in the EU and in other parts of the world (Schröter et al. 2016). Furthermore, several local experiences have proven the effectiveness of the ES approach in driving policy changes toward more sustainable outcomes in different contexts and scales (Ruckelshaus et al. 2015). Topics addressed include river basin management, climate change adaptation and mitigation, green infrastructure planning, and corporate risk management, to name just a few (Ruckelshaus et al. 2015; Dick et al. 2017), with a wide range of stakeholders involved in different decision-making processes, from landscape and urban planning (Hansen et al. 2015; Babí Almenar et al. 2018) to impact assessment (Geneletti 2016; Rozas-Vásquez et al. 2018).

The spread of the ES concept and its progressive inclusion into decision-making at various levels raised the interest on how ES and related values could be assessed in a way that allowed comparison across space and monitoring through time. Considering the type of values that they aim to capture, ES assessment methods are commonly classified in biophysical, socio-cultural, and economic methods (Harrison et al. 2017). Biophysical methods quantify ES in biophysical units based on the analysis of structural and functional traits of ecosystems, or on biophysical modelling (e.g., hydrological and ecological models, production functions). Socio-cultural methods capture individual or social preferences expressed by stakeholders in non-monetary terms (e.g., time use assessments, photo series analysis). Economic methods quantify ES values in monetary units (e.g., market prices, replacement cost, hedonic pricing). Although the distinction is sometimes blurred (e.g., methods to investigate social preference can be used to assign monetary values), it helps to understand the variety of methods from different disciplinary backgrounds that can be adopted in ES assessments (Santos-Martin et al. 2018).

While today several methods for ES mapping and assessment are well-established and have demonstrated their potential to provide useful information to decision-making (Burkhard et al. 2018), the challenge is on how multiple ES assessments can be integrated to contribute to answer real-world policy questions. On the one hand, decisions usually affect not a single but a bundle of ES (Jopke et al. 2015; Spake et al. 2017), hence assessments able to account for multiple ES and their multiple values are needed to investigate synergies and trade-offs potentially arising from decisions (Geneletti et al. 2018). On the other hand, ES assessments should be able to reflect views and opinions of the different stakeholders involved, including those that are normally under-represented (Jacobs et al. 2016). Urban planning is an example of decision-making process where complex policy questions are addressed, a broad range of stakeholders is engaged, and multiple ES values emerge. In cities, land-use decisions made during the planning process determine the availability of ES fundamental to the wellbeing of urban population. Hence, the inclusion of ES in planning is essential to promote sustainable urban development.

1.2 Planning for Ecosystem Services in Cities

Even though cities may seem to have little to do with the concept of ES, except for largely benefitting from them while threatening their provision through urbanization processes (MA 2005), this view has progressively shifted during the last years. While the ES science was developing, cities started to be seen not just as consumers of ES supplied from outside urban areas, but also as producers themselves, as already noted in the seminal work by Bolund and Hunhammar (1999). The study of urban ES, i.e. of the "ES provided by urban ecosystems and their components" (Gómez-Baggethun and Barton 2013), became a focus of ES research (Haase et al. 2014; Luederitz et al. 2015). Regulating and cultural ES emerged as the most relevant in urban areas (Gómez-Baggethun and Barton 2013; Elmqvist et al. 2016). By regulating stormwater runoff and flows, purifying the air, regulating micro-climate, reducing noise, and moderating environmental extremes, urban ecosystems affect the quality of the urban environment and control the associated hazards. Moreover, by providing suitable space for recreation, increasing the aesthetic quality of urban spaces, offering opportunities for cultural enrichment, and preserving local identity and sense of place, they provide a range of non-material benefits that are essential for human and societal wellbeing in cities (Gómez-Baggethun and Barton 2013; Elmqvist et al. 2016).

Preserving, restoring, and enhancing urban ES is therefore necessary to ensure liveable, sustainable, and resilient cities (McPhearson et al. 2015; Botzat et al. 2016; Frantzeskaki et al. 2016). Urban ES and associated benefits are linked to many of the most pressing challenges for cities. Mitigating and adapting to climate change, promoting citizens' heath, enhancing social inclusion, and reducing the environmental footprint of cities, to name just a few, all have a direct relation with the provision of urban ES (Bowler et al. 2010a; Demuzere et al. 2014; McPhearson et al. 2014). Furthermore, many urban ES produce effects only at the local level (Andersson et al. 2015) and man-made substitutes, when existing, are often characterised by high costs and impacts (Elmqvist et al. 2015). While urban population continues to grow, maintaining healthy and functioning ecosystems appears therefore of utmost importance to guarantee that the increasing demand for ES in met in a sustainable way.

Urban planning affects urban ES in multiple ways (Cortinovis 2018). First, the provision of urban ES depends on the availability and spatial distribution of urban ecosystems and their components, hence on the strategic decisions on land-use allocations that are made during urban planning processes (Langemeyer et al. 2016). Second, by defining the spatial arrangement of land uses, urban planning also determines the distribution of population and urban functions, which affects the demand for urban ES (Baró et al. 2016). Third, planning decisions also contribute to define some physical properties as well as institutional and management arrangements of the city (e.g., property type, accessibility) that play a key role in defining who can benefits from urban ES (Barbosa et al. 2007). Hence, making urban planning aware of ES and their values, and assessing the impacts of planning actions on ES provision, is fundamental to ensure that benefits from ES are preserved and enhanced.

Acknowledging the presence of nature within cities as beneficial is not an innovation in the urban planning discipline, and references to the importance of green spaces in cities and to their positive influence on the wellbeing of urban population can be traced back to the very initial stage of modern planning (see e.g. Howard 1902). However, in the last century, a view of nature in cities as only related to aesthetic and recreational values prevailed, and a strong focus on urban form as a determinant of the environmental performance of cities made other strategies, such as compactness, density, and functional diversity, prevail even when the then new paradigm of sustainability emerged (Jabareen 2006). Only recently, also thanks to a growing scientific evidence, 'greening the city' has become an imperative for urban planning. The concepts of 'ecosystem-based actions' (Geneletti and Zardo 2016; Brink et al. 2016) and 'nature-based solutions' (Raymond et al. 2017) applied to cities suggest the active promotion of urban ES and related benefits to sustainably tackle a wide range of urban challenges.

Within this framework, the integration of ES knowledge and approach in urban planning is indicated from many sides as a valuable strategy to address some of the 'wicked' problems of todays' urban development, from the necessary transition to resilience (Collier et al. 2013) to the need for sustainable approaches to address urban peripheries (Geneletti et al. 2017). That's why the inclusion of ES in urban plans started to be considered an indicator of their quality (Woodruff and BenDor 2016), ultimately measuring their capacity to put in place strategic actions towards more sustainable and resilient cities (Frantzeskaki et al. 2016).

Integrating the ES concept and approach in urban planning processes is expected to provide multiple benefits. First, to clarify the ecological – structural and functional – foundations of ES provision, thus highlighting the links between human wellbeing and the state of ecosystems (Haines-Young and Potschin 2010), hence the role of ecological knowledge in supporting effective planning actions (Schleyer et al. 2015). Second, to raise awareness on the whole range of ES and associated benefits that are produced by urban ecosystems, thus providing a comprehensive understanding of the values at stake and of the trade-offs that may arise from land-use decisions (de Groot et al. 2010). Third, to support the explicit identification of beneficiaries, including those normally under-represented in decision-making processes, thus promoting concerns for environmental justice (Ernstson 2013) and strengthening planners' arguments in balancing public and private interests (Hauck et al. 2013b).

In spite of these expectations, the integration of ES in urban planning is still limited (Cortinovis and Geneletti 2018). Haase et al. (2014), Kremer et al. (2016), and Luederitz et al. (2015) summarized the main challenges to face. Among others, they identified the need for more appropriate methods and indicators able to capture the heterogeneity and fragmentation of urban ecosystems, a scarce investigation of the relation between urban ES and biodiversity, the uncertainty about the degree of transferability of data and results, and the lack of analyses that account for ES demand by integrating people's preferences and values, particularly in the assessment of cultural ES. This book intends to contribute to address some of these challenges, promoting a full integration of the ES concept and approach in urban planning.

1.3   Book Objectives and Outline

This book analyses the integration of ES knowledge in urban planning showing and discussing how it can be promoted, to which purposes, and with what results. The overall objective is to provide a compact reference to the state of the art in this field, which can be used by researchers, practitioners, and decision makers as a source of inspiration for their activity. The books addresses the topic by: (i) investigating to what extent ES are currently included in urban plans, and discussing what is still needed to improve planning practice; (ii) illustrating how to develop ES indicators and information that can be used by urban planners to enhance plan design; (iii) demonstrating the application of ES assessments to support urban planning processes through case studies; and (iv) reflecting on criteria for addressing equity in urban planning through ES assessments that consider issues associated to supply, access, and demand of ES by citizens.

Chapters 2 and 3 review current practices, and investigate the extent to which ES are included in different types of planning instruments for cities. The ultimate objective is to understand what kind of ES knowledge is already used, and what is still needed to improve the content of plans, and their expected outcomes. In both chapters, a review framework is developed and applied to analyse the ES-related content of planning documents, irrespective of the terminology adopted. Chapter 2 focuses on urban spatial plans, and examines how nine urban ES are addressed in a sample of 22 urban plans of Italian cities. The review considers both breadth (i.e., the ES inclusion across different plan components) and depth (i.e., the quality of ES information). Chapter 3 focuses on urban climate adaptation plans, an increasingly common type of plans where ES knowledge is instrumental to inform strategies for so-called ecosystem-based adaptation (EbA) to climate change. The chapter proposes a classification of EbA measures, and reviews the extent to which they have been included in the climate-adaptation plans of 14 European cities, and the quality of the related information.

The bottlenecks of ES inclusion in current practice that emerged from Chaps. 2 and 3 set the basis to propose the way forward illustrated in the remaining of the book. Particularly, Chap. 4 presents the development of an ES model that can provide information directly usable in urban planning. The chapter focuses on microclimate regulation provided by urban green infrastructure. The model developed assesses the supply of this ES by different types of green infrastructure, relying on data that are widely available in modern urban planning practice. The application of the model is illustrated for the city of Amsterdam, The Netherlands.

Chapter 5 takes the use of ES information in urban planning a step further, by illustrating a case study where the outcomes of ES mapping and assessment are used to inform planning decisions. The micro-climate regulation model presented in Chap. 4 is applied in the city of Trento (Italy), together with a model to assess the opportunities for nature-based recreation provided by green spaces. The outcome of both models are combined with spatial information on the potential beneficiaries of the selected ES, and used to compare planning scenarios related to brownfield redevelopment. The case study demonstrates the importance of including information about ES demand and beneficiaries to understand the social implications of planning decisions, particularly in terms of equity and environmental justice.

Equity implications related to ES in urban planning are the subject of Chap. 6. This chapter identifies and discusses the key elements for analysing equity in the distribution of ES in cities, namely ES supply, access and demand. A case study application demonstrates how ES assessments should be designed, and their outcomes used, to pursue an equitable distribution of ES in cities through urban planning decisions. Finally, Chap. 7 draws come conclusions and formulate recommendations for enhancing the use of ES knowledge in planning practice.

Chapter 2
Reviewing Ecosystem Services in Urban Plans

*Text and graphics of this chapter are based on: Cortinovis C,
Geneletti D (2018) Ecosystem services in urban plans: What is
there, and what is still needed for better decisions. Land use policy
70:298–312. doi: https://doi.org/10.1016/j.landusepol.2017.10.017*

2.1 Introduction

The incorporation of ecosystem services (ES) in urban plans is considered an indicator of their quality (Woodruff and BenDor 2016) and, ultimately, of their capacity to put in place strategic actions towards more sustainable and resilient cities (Frantzeskaki et al. 2016). Using Italy as a case study, this chapter explores how urban plans integrate knowledge on ES to secure or improve ES provision by conserving, restoring, and enhancing urban ecosystems. The ultimate objective is to shed light on what ES information is already included in current urban plans to support planning actions, and what is still needed to improve their content and decisions.

Scientists have monitored the uptake of ES in planning practices mainly following two approaches. The first approach investigates how practitioners, policy-makers, and stakeholders understand the concept of ES. Perceived opportunities and limitations in the use of ES in planning are usually elicited from key informants through interviews (see examples in Beery et al. (2016); Hauck et al. (2013a); Niemelä et al. (2010)). The results of similar studies are useful to understand the mechanisms through which the uptake of ES can occur. However, being based on self-reported perceptions and opinions, these studies do not measure the actual level of implementation of the ES concept into planning practices. The second approach reviews the content of documents, including strategic plans (Piwowarczyk et al. 2013), environmental policies (Bauler and Pipart 2013; Maczka et al. 2016), and urban plans (Hansen et al. 2015; Kabisch 2015) using content or keyword analysis.

Investigating the uptake of ES as a new planning paradigm may lead to overlook the fact that urban plans have a tradition of accounting for - at least some – ES. ES-inclusive approaches have routinely been used in planning, even though under different names, as it clearly emerges from both planners opinions (Beery et al. 2016), and historical analyses of planning documents (Wilkinson et al. 2013). To understand how the ES approach can contribute to improve the current planning practices, it is necessary to identify which urban ES are addressed and how, and to what extent the conceptual framework of ES is already integrated in urban plans. To this aim, this chapter investigates the contents of plans by searching for explicit but also implicit references to ES, and classifying the information based on their use within the plan, as described in the next Section.

2.2   Methods to Analyse ES Inclusion in Urban Plans

We selected a sample of 22 recent urban plans of Italian cities (see Annex 1). Urban plans in Italy are comprehensive spatial planning documents drafted at the municipal level, fairly similar in content to analogous documents around the world. Their main tasks are: defining land-use zoning; designing and coordinating the system of public spaces and public services; detailing and integrating regulations and provisions set by higher administrative levels. The plans were analysed through a directed qualitative content analysis composed of the three steps described next.

2.2.1   Assessing the Breadth of Inclusion

We considered the following urban ES: food supply, water flow regulation and run-off mitigation, urban temperature regulation, noise reduction, air purification, moderation of environmental extremes, waste treatment, climate regulation, and recreation. Following previous content analyses of urban plans (Geneletti and Zardo 2016; Woodruff and BenDor 2016), we identified three main plan components: information base, vision and objectives, and actions. The *information base* component illustrates the background knowledge that supports planning decisions. The *vision and objectives* component states the long-term vision of the plan and the targets that the plan pursues. The *actions* component illustrates decisions taken by the plan, including strategies and policies (projects, regulations, etc.) that are envisioned to achieve the objectives. Urban ES and plan components are cross-tabulated in a table, which is filled for each plan under investigation by analysing both its textual and cartographic documents, and reporting the relevant content. The number of filled cells in the table allows measuring the overall breadth of inclusion of the analysed ES. We adopted the formulation of the breadth score indicator proposed by Tang et al. (2010) and later applied by Kumar and Geneletti (2015). We calculated the breadth score both for the whole plans and for each component individually.

2.2.2 Assessing the Quality of Inclusion

Quality is conceptualized as the presence of desired characteristics, described through criteria that high-quality plans are expected to meet (Berke and Godschalk 2009). We built on the scoring protocol developed by Baker et al. (2012), and adopted a 5-point scale, with scores ranging from 0 (no inclusion) to 4 (high-quality inclusion). A plan is awarded the highest score in the *information base* component when it acknowledges the links between ecosystems and human wellbeing, identifies functions and processes that determine the provision of ES, and applies this knowledge to a quantitative assessment of the local provision that also includes an analysis of demand and beneficiaries (Table 2.1).

Table 2.1 Scoring protocol for the *information base* component. The examples are taken from the analysed plans (own translation). Plan ID codes are reported in Annex 1

Score	Description	Example
0	The plan contains no evidence of the ES concept.	–
1	The plan acknowledges the link between ecosystems and ES supply, either explicitly as part of the information base, or implicitly in the description of objectives and actions.	"Urban green areas […] guarantee protection of biodiversity inside the city as well as recreation and compensation of anthropogenic impacts." [explicit] (Source: P12)
		"Acoustic green belts with a minimum length of 50 m […] must be composed of evergreen broadleaves hedges or trees, with preference for fast growing, indigenous species with large crowns". [implicit in the description of actions] (Source: P21)
2	The plan mentions functions and processes on which ES provision depends, and identifies the elements that define ES potential. However, it lacks local application and analysis.	"Urban micro-climate […] can be enhanced by the presence of vegetation […]. A continuous green network that crosses the city, linked to the countryside, constitutes a ventilation corridor that enhance urban micro-climate. The most relevant biophysical process that determines the effects of vegetation on urban climate is the transpiration (…)". (Source: P06)
3	The plan shows a limited level of locally specific application of the ES concept. A basic qualitative assessment of the current state of ES is performed, but detailed analysis, quantitative measurements, and clear identification of demand and beneficiaries are lacking.	"Land-use changes determine an increase in soil sealing with higher storm water run-off. […] The increase in soil sealing and, consequently, in the flow rates produced by the reference rain event were quantified based on the distribution of sealed surfaces (e.g. streets, roofs) and permeable surfaces (e.g. parks) in each transformation area, as proposed by the draft masterplan". (Source: P20)
4	The plan shows an in-depth application of the ES concept in the analysis of the local provision of urban ES, including quantitative measurements, detailed assessment, and identification of demand and beneficiaries.	Spatially explicit mapping of the accessibility to recreational areas (5 classes of accessibility), and quantification of beneficiaries broken down by age group (< 3; between 4 and 7; between 8 and 14; > 64 years). (Source: P04)

Table 2.2 presents the scoring protocol used for the *vision and objectives* component. A plan is awarded the highest score when it defines locally specific principles and quantitative targets for the enhancement of ES provision. A high-quality *vision and objectives* component is expected to coordinate public and private land-use decisions to achieve the defined goals (Berke and Godschalk 2009), and, more specifically, to guide the choice of the best planning alternatives in terms of both "what" and "where" (Kremer and Hamstead 2016). For the *actions* component, we assigned a binary score to record the presence, for each urban ES, of at least one action (as in Wilkinson et al. (2013)). We then defined the overall quality of the component as the share of ES addressed by at least one action in the plan. To measure the overall quality of inclusion in the sample, we adopted the depth indicator proposed by Tang et al. (2010), which calculates the average score considering only the plans with a non-zero score in the component. We calculated the indicator for each urban ES for the *information base* and for the *vision and objectives* components.

Table 2.2 Scoring protocol for the *vision and objectives* component. The examples are taken from the analysed plans (own translation). Plan ID codes are reported in Annex 1

Score	Description	Example
0	The plan contains no evidence of objectives related to the ES.	–
1	The plan defines objectives of ecosystem conservation/ enhancement, which are expected to affect positively ES provision, but does not directly refer to ES.	"Allow the restoration of river sides, particularly of potential flooding risk areas and retention areas that control overflows". (Source: P11)
2	The plan defines objectives directly related to ES provision. However, they are entirely descriptive, and lack local application and analysis.	"Tree planting, enlargement of existing green areas, and hedge planting must be encouraged to enhance the local micro-climate (including air purification, noise abatement, and mitigation of the heat island caused by impermeable surfaces)". (Source: P07)
3	The plan defines qualitative objectives directly related to ES provision through a locally specific analysis and application of the ES concept.	[In the peri-urban areas] "the municipal administration envisions the drafting of a specific plan […] for the safeguard and enhancement of green recreational areas and green belts, aimed at increasing the absorption of particulate matter and the reduction of the urban heat island effect." (Source: P10)
4	The plan defines objectives and quantitative targets related to ES provision through a locally specific analysis and application of the ES concept.	"The objective of increasing the amount of public green areas up to three times the existing can also be reached by making the 22% of the actual inaccessible green areas accessible and usable. This way, the green area per inhabitant doubles and exceeds the 30 Km²/inhabit..". (Source: P09)

2.2.3   Analysing Planning Actions

We investigated three action properties, namely typology, target area, and implementation tool. The typology describes the type of intervention on urban ecosystems, i.e. conservation, restoration, enhancement, or new ecosystem. The target area describes the scale of the planning action and the spatial distribution of the interventions within the city, i.e. widespread over the whole territory, targeting specific areas, or limited to specific sites. The implementation tool describes the type of legal instruments provided to implement the action, i.e. regulatory tools, design-based tools, incentive-based tools, land acquisition programs, or other tools (Table 2.3). A list of planning actions addressing each of the nine urban ES was compiled for each plan. Then, actions were classified with respect to the three properties, and recurrent combinations were identified both in the whole sample and for each urban ES.

Table 2.3 Categories and sub-categories adopted for classifying planning action properties

Typology	Description
Conservation	Action aimed at preserving the current state of urban ecosystems in order to secure the provision of ES. (*e.g. preserving existing wetlands*)
Restoration	Action aimed at recovering the health and functionality of urban ecosystems in order to get back to a level of ES provision offered in the past. (*e.g. de-paving sealed surfaces*)
Enhancement	Action aimed at improving the state of existing urban ecosystems in order to enhance the provision of ES. (*e.g. enlarging existing urban parks*)
New ecosystem	Action aimed at creating new urban ecosystems in order to provide new ES in an area. (*e.g. planting street trees*)
Target area	**Description**
Widespread	The action targets all the future interventions of a certain typology. (*e.g. new building interventions, demolitions and reconstructions, large urban transformations*)
Specific areas	The action targets one or more zones in which the plan divides the city, or areas in the city identified by the presence of a specific issue. (*e.g. industrial sites, agricultural fragments*)
Specific sites	The action targets a specific project site or transformation area envisioned by the plan (*e.g. a specific urban park, a specific brownfield to be re-developed*)
Implementation tool	**Description**
Regulatory tools	
Building code standard or requirement	Definition of a standard or a requirement in the building code that must be met when developing or re-developing an area.
Compensation measure	Definition of a compensation measure (e.g. payments for realizations, mandatory land property transfers), including its rationale and quantification.

(continued)

Table 2.3 (continued)

Conservation zone or protected area	Definition of a boundary for a conservation zone or a protected area, and of the rules (restrictions and limitations) that must be respected within this area.
Other regulatory tools	All the other types of actions undertaken through regulatory tools (e.g. density regulations, permitted and forbidden uses related to zoning).
Design-based tools	Definition of specific design solutions to implement either in public projects or in privately lead urban developments.
Incentive-based tools	
Preferential tax treatment	Definition of a financial incentive in the form of a preferential tax treatment (usually a reduction in planning fees).
Density bonus	Definition of a non-financial incentive in the form of an increase in the surface (or volume) that is allowed in the area.
Transfer of development rights	Definition of a "transfer of development rights" mechanism: the development right is assigned to an area as a compensation for the placement of a conservation easement that prevents further development, and can be applied in other areas or sold. Participation is on a voluntary basis.
Other incentive-based tools	This category includes all the other types of incentive-based tools, such as the possibility of realizing specific interventions under certain conditions.
Land acquisition programs	Definition of a program for land acquisition by the public administration, with the aim of realizing a public project.
Other tools	
Principles for public space design	Definition of design principles and guidelines (non-compulsory) that should be applied in the realization of public spaces.
Principles for territorial management	Declaration of principles that the municipal administration will follow in the management of the territory (e.g. commitment in administrative processes or in the implementation of future planning documents). It also includes assessment criteria for proposed interventions, when no incentive is envisioned.
Promotion of good practices	Suggestion of principles, good practices, best available techniques, etc. (non-compulsory) to apply in private areas.

2.3 Results

2.3.1 Breadth of ES Inclusion in Urban Plans

Figure 2.1 shows the breadth score indicator measuring the overall inclusion in plans (i.e. inclusion in at least one component). Urban ES are clearly divided into two groups: five urban ES are included in almost all plans in the sample (breadth score > 85%), whereas around half of the plans consider the other four urban ES (breadth score between 45% and 55%). Figure 2.2 breaks down the breadth score by plan component. The frequency of mention in the *information base* and in the *actions* components is similar across ES, although values for the latter are slightly

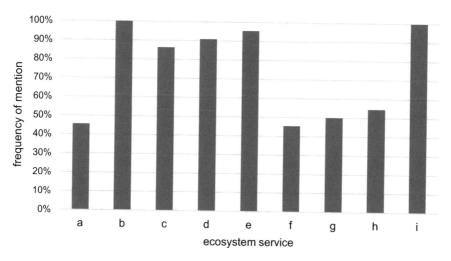

Fig. 2.1 Breadth score indicator measuring the inclusion of urban ES in at least one component of plans. ES are named as follows: (a) food supply, (b) water flow regulation and runoff mitigation, (c) urban temperature regulation, (d) noise reduction, (e) air purification, (f) moderation of environmental extremes, (g) waste treatment, (h) climate regulation, (i) recreation

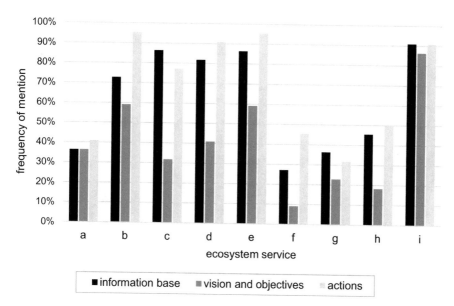

Fig. 2.2 Breadth score indicator measuring the inclusion of urban ES in the three plan components. ES are named as follows: (a) food supply, (b) water flow regulation and runoff mitigation, (c) urban temperature regulation, (d) noise reduction, (e) air purification, (f) moderation of environmental extremes, (g) waste treatment, (h) climate regulation, (i) recreation

higher. The frequency of mention in the *vision and objectives* component is generally lower, with the only two exceptions of food supply and recreation, which are mentioned evenly in the three components.

2.3.2 Quality of ES Inclusion in Urban Plans

The overall quality of ES inclusion (Fig. 2.3) is generally low, with only two plans in the sample reaching the score of 1.5 in the 0–3 range obtained by summing the normalized scores in the three components. The *actions* component receives the highest average normalized score (0.65), while normalized scores for the *information base* and the *vision and objectives* components are lower than 0.5 in all plans. When looking at the distribution of quality scores for the different urban ES in the different plan components, it emerges that the most common quality score in the *information base* component is equal to 1. However, the same pattern discussed for the breadth indicator emerge with respect to the different ES. Although the overall performance is quite poor, five ES (water flow regulation and runoff mitigation, recreation, air purification, noise reduction, and urban temperature regulation) are addressed in this component more often and with a higher quality compared to the others. Water flow regulation and run-off mitigation and recreation are the only ones for which some of the plans were given the highest scores. However, only analyses

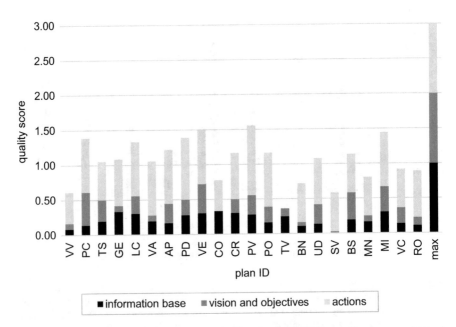

Fig. 2.3 Overall quality of ES inclusion calculated as the sum of the normalized scores obtained in the three components. Plan IDs can be found in Annex I

of recreation show, in some cases (around 30%), consideration for demand and beneficiaries. In the *vision and objectives* component, the pattern is less clear. Here, the most common quality score is 0, which indicates the absence of any reference to ES. However, the highest scores (3 and 4) are more frequent than in the *information base* component, and are found at least in one plan for almost all ES, even though a quality score of 4 is again obtained only by water flow regulation and runoff mitigation and recreation. The depth score indicator (Fig. 2.4) confirms that, when ES are considered, the average quality of the *vision and objectives* component is higher compared to the *information base* component.

2.3.3 Actions Related to ES in Urban Plans

In total, 526 actions addressing urban ES were identified, distributed as shown in Fig. 2.5. Recreation is by far the most commonly address ES, with an average of more than eight actions per plan. An average of three to four actions per plan address water flow regulation and runoff mitigation, noise reduction, and air purification, with implicit acknowledgement of the demand for mitigation of these common urban environmental problems. The other services are addressed on average by less than two actions per plan. Table 2.4 lists the most frequent actions for each urban ES, based on the type of intervention proposed.

Figure 2.6 describes the distribution of actions according to the three properties (typology, target area, and implementation tool). New interventions, such as the

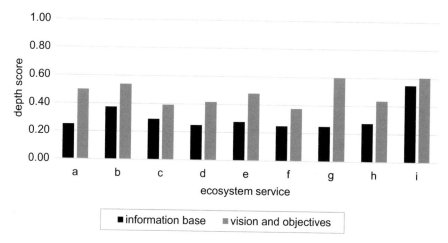

Fig. 2.4 Depth score indicator measuring the quality of inclusion of urban ES in the *information base* and in the *vision and objectives* components. ES are named as follows: (a) food supply, (b) water flow regulation and runoff mitigation, (c) urban temperature regulation, (d) noise reduction, (e) air purification, (f) moderation of environmental extremes, (g) waste treatment, (h) climate regulation, (i) recreation

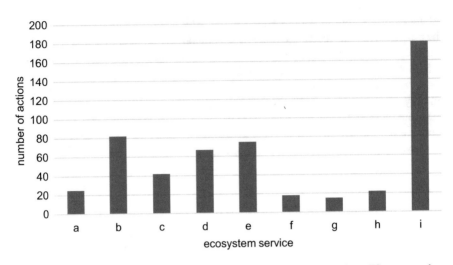

Fig. 2.5 Number of actions addressing each ES in the whole sample of plans. ES are named as follows: (a) food supply,(b) water flow regulation and runoff mitigation, (c) urban temperature regulation, (d) noise reduction, (e) air purification, (f) moderation of environmental extremes, (g) waste treatment, (h) climate regulation, (i) recreation

Table 2.4 Groups of actions based on the type of intervention proposed. Only actions recurring in more than three plans are reported

Urban ES and related actions	Number of plans
Food supply	
Realization of new allotment gardens	6
Conservation of existing allotment gardens and residual agricultural patches	4
Water flow regulation and runoff mitigation	
Prescription of a minimum share of unsealed surfaces to maintain in new developments	14
Prescription of permeable pavements for parking areas, cycling paths, etc.	9
Realization of green roofs	6
Realization of bio-retention basins or other ecosystem-based approaches to storm-water management	6
De-paving	5
Urban temperature regulation	
Provision of trees to shade parking areas	10
Creation of new green areas/enlargement of existing green areas	7
Noise reduction	
Realization of green barriers/areas for noise shielding from infrastructures	15
Realization of green barriers/areas for noise shielding from factories and plants	15
Soil modeling for noise protection	4
Generic use of green for noise shielding	4

(continued)

Table 2.4 (continued)

Urban ES and related actions	Number of plans
Air purification	
Realization of green barriers/areas for air purification from traffic emissions	15
Realization of green barriers/areas for air purification from industrial emissions	13
Creation of woodlands and urban forests	5
Generic use of green for air purification	4
Conservation of existing green areas	4
Realization of green roofs and green walls	4
Moderation of environmental extremes	
Enlargement of river areas and conservation/reclamation of floodplains	8
Waste treatment	
Climate regulation	
Realization of Kyoto-forests and new woodlands	8
Increase of public green areas	5
Recreation	
Realization of new public green spaces and urban parks	16
Strengthening walking and cycling accessibility among green areas and with the rest of the city	16
Increasing fruition of green spaces through new walking and cycling paths	14
Restoration of existing green areas aimed at increasing their use	14
Promotion of new functions and uses in the existing green spaces	12
Enlargement of existing green spaces	8
Identification of opportunities for recreation in agricultural areas	8
Realization of peri-urban parks	7
Opening of existing private/unused gardens and green spaces to public use	6

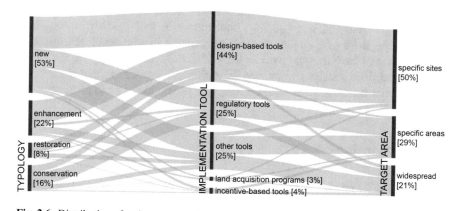

Fig. 2.6 Distribution of actions per typology, target area, and implementation tool, and recurring combinations in the whole sample of actions

realization of new green areas, represent the most common typology of action (53%). Around 44% of the actions rely on design-based implementation tools (e.g. projects included in the plan), through which the public administration can control action implementation with a quite high level of detail. Regulatory tools, particularly the definition of standards and other specific requirements in building codes, and other tools, such as the suggestion of good practices, are also among the most common, both with 25% of the sample. Incentive-based tools (e.g. density bonuses) and land acquisition programs are the least adopted tools, and accounts for only 4% and 3% respectively. In terms of target areas, specific sites are the most common and represent the target of 50% of the actions. These include, for example, the restoration of specific ecosystems, the identification of conservation areas, and the realization of new urban parks. Around 29% of the actions target specific areas in the municipal territory, such as regulations to be applied in industrial areas or safeguards to protect agricultural patches. Finally, 21% of the actions are widespread. These include requirements for all new building interventions and rules to respect in case of demolitions and reconstruction.

Actions on specific sites are usually implemented through design-based tools, while actions on specific areas are generally implemented through regulatory tools or other "soft" tools such as the suggestion of good practices. Soft tools also clearly prevail in the case of widespread measures. Concerning typologies, conservation actions are more often implemented through regulatory tools, while for both enhancement and restoration activities the preferred tools are design-based. For example, new conservation areas are often defined through a boundary in the maps and a set of rules, while restoration measures are often proposed through a more detailed design.

When looking at individual ES, conservation actions are the preferred typology for improving food supply (conservation of agricultural patches) and water flow regulation and runoff mitigation (conservation of existing unsealed surfaces). Recreation is mostly promoted through enhancement interventions on existing green and blue areas. Water flow regulation and runoff mitigation also differs in term of target areas and, consequently, implementation tools, mostly prescriptions related to the share of unsealed surfaces to maintain in new developments. Two other ES do not have design-based as the preferred tools: food supply, for which 40% of the actions consist in principles for territorial management, and waste treatment, which is commonly addressed through the promotion of good practices.

2.4 Conclusions

Our review of 22 urban plans focused on the use of the ES concept as a tool to support decision-making (Mckenzie et al. 2014), as opposed to the explicit uptake of the term "ecosystem services". Similarly to what has been observed for the concept of sustainable development (Persson 2013), our hypothesis was that an effective integration should build on what is already there, and follow a mechanism of

Table 2.5 Summary of the main findings

What is already there	What is still needed
Urban planning addresses urban ES through a high number and a great variety of actions	Scientific knowledge is only partly transferred to planning practices
A wide range of local problems can be addressed through ES-based actions	There is little guidance on how to incorporate information on ES into planning processes
Urban planners are already equipped with a large set of tools to implement ES-related actions	Usable methods to assess urban ES at a relevant scale while accounting for multi-functionality of ecosystems are still lacking
Recreation provided by urban ecosystems, although not linked to the ES concept, is widely acknowledged and promoted by planning actions	Plans contain no analyses of ES demand and of the existing and expected beneficiaries (with the only exception of recreation)
A set of key regulating ES to address pressing urban environmental problems (i.e. water flow regulation and runoff mitigation, air purification, urban temperature regulation, and noise reduction) are widely acknowledged and addressed	ES are not considered a strategic issue in urban planning

"internalization" that does not necessarily require rethinking or reshaping current practices. Our findings, summarized in Table 2.5, reveal that current urban plans already include a high number of ES-related actions and a variety of tools for their implementation. This indicates that planners have the capacity and the instruments to enhance the future provision of urban ES. Actions in the analysed plans often go beyond those ordinarily mentioned as good practices, and the range of issues that they address is wider. This demonstrates a certain level of creativity that, combined with traditional ecological knowledge and the understanding of local social-ecological systems, enables the design of locally relevant interventions.

However, our study unveils a two-speed integration of urban ES, with a set of services that are widely addressed by urban plans (recreation, above all, but also regulating services linked to environmental problems typical of urban areas), and others that are hardly considered. The least considered (e.g. waste treatment and moderation of environmental extremes) are also the least popular in the scientific literature (Haase et al. 2014), and when they are included in urban plans, their treatment is very shallow (e.g. suggestion of one-fits-all good practices). This can be ascribed, at least partly, to gaps in the scientific literature, which has not produced methods and guidance that fit urban planning practices.

A further understanding and appropriation of the ES approach by urban planning would benefit future practices in many respects. First, it could promote consideration of a larger set of urban ES, at least in the initial phases of planning processes, thus increasing awareness of all values at stake, highlighting co-benefits and trade-offs that may arise from planning actions, and making prioritization more transparent. Second, it could strengthen the consideration of ES as a strategic issue for urban planning, thus promoting the definition of objectives and targets for ES enhancement, and ensuring long-term commitment in the implementation and monitoring of planning actions. Finally, it could support the explicit identification of ES demand and beneficiaries, thus improving baseline information to address urban environmental equity, and providing planners and decision-makers with stronger arguments against conflicting interests on land-use decisions.

Chapter 3
Reviewing Ecosystem Services in Urban Climate Adaptation Plans

Text and graphics of this chapter are based on: Geneletti D, Zardo L (2016) Ecosystem-based adaptation in cities: An analysis of European urban climate adaptation plans. Land use policy 50:38–47. doi: https://doi.org/10.1016/j.landusepol.2015.09.003

3.1 Introduction

In this chapter, we focus on one specific type of urban planning instrument, which has become increasingly common in the last years: urban climate adaptation plans. In these plans, ecosystem service (ES) knowledge is instrumental to propose strategies for ecosystem-based adaptation (EbA) to climate change. EbA is defined as the use of biodiversity and ES to help people to adapt to the adverse effects of climate change. EbA approaches include management, conservation and restoration of ecosystems that, by delivering ES, can help to reduce climate change exposure and effects (Munang et al. 2013). EbA can play an important role in urban contexts and help to cope with increased temperature, flood events, and water scarcity by reducing soil sealing, mitigating the heat island effect, and enhancing water storage capacity in urban watersheds (Gill et al. 2007; Grimsditch 2011; Müller et al. 2014).

The recent literature has addressed the potential role of EbA in cities (Berndtsson 2010; Bowler et al. 2010b; Müller et al. 2014). In particular, Demuzere et al. (2014) presented a comprehensive analysis of the available empirical evidence about the contribution of green infrastructures to climate change adaptation in urban areas. Nevertheless, the concept of EbA is still relatively new for cities, and little evidence is available on the inclusion of EbA measures in actual urban plans and policies (Wamsler et al. 2014). Urban planning, at least in more industrialized countries, has been increasingly addressing climate adaptation strategies and actions, as shown by recent reviews of planning documents undertaken for cities in Europe (Reckien et al. 2014), the UK (Heidrich et al. 2013), Australia (Baker et al. 2012) and North America (Zimmerman and Faris 2011). However, none of these papers addresses specifically EbA.

In this chapter, we develop a framework to analyse the inclusion of EbA in urban climate adaptation planning, and apply it to a sample of plans in Europe. Specifically, we aim at answering the following questions:

– What are the most common EbA measures found in urban climate adaptation plans? To what climate change impact do they respond?
– In what parts of the planning documents are EbA measures present? How well and how consistently are they treated?

The ultimate purpose of the chapter is to provide an overview of the current state of the art related to the inclusion of ES in urban climate planning through EbA, and use it to identify and discuss the main shortcoming and propose possible solutions.

3.2   Methods to Analyse Urban Climate Adaptation Plans

We focused on a sample of cities considered active in climate change adaptation, by referring to the "C-40" initiative. The C-40 was established in 2005 as a network of large cities worldwide that are taking action to reduce greenhouse gas emissions and to face climate risks. This sample offers the advantage of providing information on different initiatives undertaken by cities that have been particularly active in climate adaptation strategies. Among the cities of the C-40 database, we selected the ones belonging to Member States of the European Union. We then gathered all the urban climate change responses in the form of planning documents approved by the relevant municipal authority and available on the internet (see Annex II). We use the term 'climate adaptation plan' to refer in general to plans that include strategies to reduce vulnerability to climate change in cities, even though the actual name of the plan might be different.

3.2.1   Classification of EbA Measures

As a first step, we identified and classified possible measures for EbA that are relevant for urban areas. The list of EbA proposed by EEA (2012) was revised and integrated with other typologies found in the literature. This resulted in the classification presented in Table 3.1, where definition, rationale and supporting references are provided for each measure. Measures are associated to the main climate change impact they are meant to reduce, even though it is recognized that synergies occur. For example, green roofs may contribute to reduce runoff water quantity (Berndtsson 2010), in addition to contributing to micro-climate regulation through cooling. EbA measures play at different spatial scales, ranging from building-scale interventions (e.g., green roofs and walls) to urban-scale interventions (e.g., citywide green corridors). Despite their difference in scale, the identified measures are all within the scope of urban plans; hence, they can be (at least partly) implemented by actions

Table 3.1 The classification of EbA measures for urban areas adopted in this research (building on the list proposed by EEA 2012)

EbA Measure	Climate change impact	Rationale	References
(a) Ensuring ventilation from cooler areas outside the city through waterway and green areas	Heat	If carefully designed, urban waterways and open green areas have the potential to create air circulation and provide downwind cooling effect.	Oke (1988)
(b) Promoting green walls and roofs	Heat	Vegetated roofs and facades improve the thermal comfort of buildings, particularly in hot and dry climate	Bowler et al. (2010b); Castleton et al. (2010); Skelhorn et al. (2014)
(c) Maintaining/enhancing urban green (e.g., ecological corridors, trees, gardens)	Heat	Green urban areas reduce air and surface temperature by providing shading and enhancing evapotranspiration. This cooling impact is reflected, to some extent, also in the building environment surrounding green areas.	Yu and Hien (2006); Demuzere et al. (2014)
(d) Avoiding/reducing impervious surfaces	Flooding	Interventions to reduce impervious surfaces in urban environments (e.g., porous paving; green parking lots; brownfield restoration) contribute to slow down water runoff and enhance water infiltration, reducing peak discharge and offering protection against extreme precipitation events.	Jacobson (2011); Farrugia et al. (2013)
(e) Re-naturalizing river systems	Flooding	Restoring river and flood-plain systems to a more natural state in order to create space for floodwater can support higher base flows, reducing flood risk. Restoration interventions include, for example, the establishment of backwaters and channel features and the creation of more natural bank profiles and meanders.	Palmer et al. (2009); Burns et al. (2012)
(f) Maintaining and managing green areas for flood retention and water storage	Flooding, Water scarcity	Vegetated areas reduce peak discharge, increase infiltration and induce the replenishment of groundwater. To enhance this, retention basins, swales, and wet detention systems can be designed into open spaces and urban parks.	Cameron et al. (2012); Liu et al. (2014)
(g) Promoting the use of vegetation adapted to local climate and drought conditions and ensuring sustainable watering of green space	Water scarcity	Green space may exacerbate water scarcity in urban areas. To limit this problem, interventions can be directed at choosing the most appropriate tree species (that are drought resistant but still suitable as a part of the urban green space), and designing sustainable watering systems (e.g., using grey water or harvested rainwater)	EEA (2012)

proposed in planning instruments. Measures such as river re-naturalization, in most cases, cannot be handled within the border of a city alone. However, urban plans have the possibility to implement these interventions (at least for the urban sector of rivers), as well as to promote coordination with other planning levels (e.g., regional planning, river basin planning). Thus, these measures have been included in the proposed classification of EbA measures relevant for urban areas.

3.2.2 Analysis of the Content of the Plans

As in Chap. 2, the content of the plans was divided into different components, which represent thematically different parts of the plans. For climate adaptation plan, four components were identified: *information base*; *vision and objectives*; *actions*; *implementation*. The *information base* includes the analysis of current conditions and future trends (typically presented in the introductory parts of the planning documents), which is performed in order to provide a basis for the subsequent development of the plan's objectives and actions. *Vision and objectives* include the statement of the ambition and of the general and specific objectives that a plan intends to achieve. *Actions* include all the decisions, strategies and policies that the plan propose, in order to achieve its objectives. Finally, *implementation* refer to all measures (including budget-related ones) proposed to ensure that actions are carried out.

Similarly to the previous Chapter, a direct content analysis was performed, by reading all the documents associated to the selected plans and identifying – for each of the four components - the content related to EbA measures, using the classification presented in Table 3.1. This approach was preferred to a keyword-based analysis, given that there is not yet a well-established terminology in this field, and plans use a wide range of different wording to refer to concepts related to EbA and to ES in general (Braat and de Groot 2012). Hence, we searched for the presence of the different measures, irrespective of whether the plan used the term "EbA" or not to describe them.

The content analysis followed a two-step process. First, the presence of the different EbA measures in each plan component was searched, by using the following guiding questions:

– *Information base*: Does it contain data/statements/analyses that show awareness about EbA?
– *Vision and objectives*: Are there objectives associated to the development/ enhancement of EbA measures?
– *Actions*: Are there actions aimed at developing/enhancing EbA measures?
– *Implementation*: Do the implementation provisions include reference to EbA measures?

Second, whenever the answer to the previous questions was positive, the content was further analysed in order to assess the extent to which EbA measures were addressed, by using the four-level scoring system presented in Table 3.2. Finally, an

Table 3.2 Scoring system used to evaluate the plan components

Score	Information base	Vision and objectives	Actions	Implementation
0	No evidence of information related to EbA measures	No evidence of objectives related to EbA measures	No evidence of EbA measures	No evidence of implementation provisions related to EbA measures
1	Acknowledges EbA measures only generally (not in connection to specific climate change issues)	Mentions EbA-related objectives, but lacks further definition	Mentions EbA measures, but lacks further definition	Mentions implementation provisions related to EbA measures, but lacks further definition
2	Acknowledges EbA measures in the context of specific climate change issues	Includes EbA measures in the objectives and provides some details on their specific content and how to pursue them	Includes EbA measures in the actions and provides some details on their application and activities	Includes EbA-related implementation provisions and provides some details on their application
3	Acknowledges EbA measures and describes (at least qualitatively) the potential climate change adaptation effects	Includes EbA measures in the objectives, provides details on their content, and describes links with related planning and policy processes at the local/regional level	Includes EbA measures in the actions, provides information on their application and activities, including locally-specific details	Includes EbA-related implementation provisions and provides information on their application, including details on budget, responsible bodies, etc.

average score was obtained for each type of EbA measure by computing the average value obtained by that measure in all the plans where the measure is found, and for all plan components.

3.3 Results

3.3.1 What EbA Measures are Included in the Plans and How?

In total, 44 EbA measures were found in the selected plans. Figure 3.1 illustrates the breakdown in the seven types. As can be seen, measures *c* (maintaining/enhancing urban green) and *f* (maintaining and managing green areas for flood retention and water storage) are the most common ones, and are found in 85% of the selected plans. Examples of measures *c* include efforts to increase green areas and neighbourhood gardens (Paris), proposals for enhancing the connectivity among existing

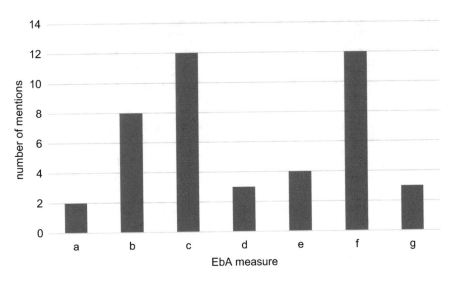

Fig. 3.1 Number of mentions of the seven types of EbA measures (see legend in Table 3.1) in the sample of plans

green areas through the design of green corridors and rings (Milan), and the use of plants to provide shade in new industrial estates (Amsterdam). Measures *f* consist, for example, in the creation of new wetland areas and ponds (Berlin), and the design of green spaces to store rainwater in the event of torrential rain (Copenhagen).

Measure *b* (promoting green walls and roofs) was found in 57% of the plans. For example, Paris's plan contains provisions for the establishment of roof and wall gardens (measure *b*), including the identification of priority spots for this type of green infrastructures. Measure *e* (re-naturalizing river systems) was found in 29% of the plans. In Madrid, for example, this consisted in a series of bank improvement projects aimed at reducing flood hazard and expanding riverside public space. Measures *a*, *d* and *g* (respectively, ensuring ventilation, avoiding/reducing impervious surfaces, and promoting climate-adapted vegetation and sustainable watering) were less common, and found only in 14–21% of the plans. For example, concerning measure *a*, cold air networks to ensure ventilation and prevent over-heating are mentioned in Copenhagen's plan, whereas Madrid's provides for the promotion of *ecobarrios* where ventilation will be one of the factors considered in the design of greening interventions. Berlin's plan attains the reduction of impervious surfaces (measure *d*) through renovation projects for buildings and school playgrounds that include interventions to improve soil permeability and in situ infiltration. Finally, concerning measure *g*, Venice's plan promotes the use of autochthonous species adapted to the local climate, and Madrid's contains detailed guidelines for "sustainable gardens" with recommendations for the selection of plant species and sustainable watering systems.

The results of the application of the scoring systems were used to compute an average score for each type of EbA measure (Fig. 3.2), representing the average

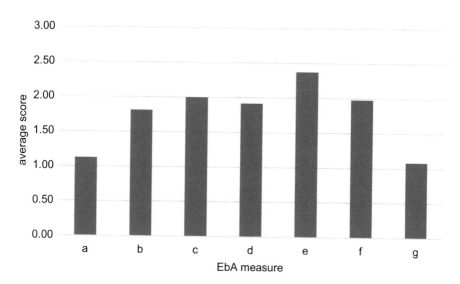

Fig. 3.2 Average scores of the seven types of EbA measures (see legend in Table 3.1)

value obtained by the measure in all the plans where it is found, and for all plan components. The average score ranges from 1.1 (achieved by measures *a* and *g*) to 2.4 (measures *e*). Measures *c* and *f*, which are the most frequently found, are also the ones with the highest scores, together with action *e*.

3.3.2 How Are EbA Measures Reflected Within Plan Components?

Figure 3.3 shows in which plan components EbA measures are reflected. About 91% of the measures are present in the *vision and objectives* component. This means that, when a plan includes an EbA measure, this is very often listed as (part of) one of the objectives that the plan intends to achieve. For example, Paris's plan objectives include the development of a multi-year scheme to promote roof gardens. Almost 91% of the EbA measures are addressed in the *actions* component, meaning that the plans include specific policies or activities to attain them. For example, Milan's plan includes a series of linear greening interventions along canal banks, roads, biking routes, etc. The *information base* component of the plans contains data relevant to EbA measures only in 79% of the cases. That is, 21% of the measures found in the plans are not supported by any baseline information or analysis. Even when baseline information is present, this consists mostly of general statements and descriptions. For example, Berlin's plan contains descriptions of how energy efficiency of buildings or industry could be usefully combined with projects to support sustainable local water management systems, by increasing the permeability of soil and planting

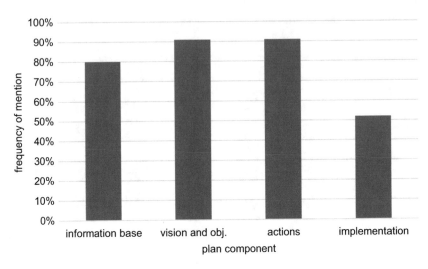

Fig. 3.3 Frequency of presence of information about the 44 EbA in the different plan components

vegetation. The *implementation* component of the plans performs even more poorly: references to EbA measures are found in only 52% of the cases. Therefore, about half of EbA measures are not associated to any action to ensure that they are carried out. When information about implementation measures are present, this consists mainly of budget-related details, as for example in the case of Madrid's plan (where each action is linked to a plan of implementation and budget), and Rotterdam's, where there are indications about green roofs subsidies.

In order to assess how well EbA measures are reflected within the different plan components, we computed the average score obtained by all EbA measures that are found in each of the four components. For example, out of the 44 EbA measures, 35 are present in the *information base* component of the selected plans. The average score represents the average of the scores obtained by these 35 EbA according to the adopted scoring system. The results show that *actions* component scored the highest (average score: 2.8), followed by the *implementation* (2.5), the *vision and objectives* (2.2) and the *information base* (1.8). Concerning the good performance of *actions*, examples include London's plan, which describes in detail the actions and associated sub-actions, specifies the responsible bodies and identifies links with other plans and policies. Similarly, Madrid's plan provides action fact-sheets, with the identification of responsible bodies and associated budget. The poorer scores of the *visions and objectives* component are because their description tend to be very general. The *information base* typically lacks details on the links between measures and climate-related issues, particularly concerning the results expected from the application of the measure.

Finally, Fig. 3.4 provides a visual overview of the distribution of information on the identified EbA measures across plan components. This figure helps to understand how consistently EbA measures are treated across the different plan components, and

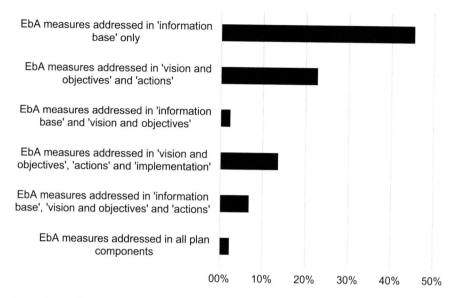

Fig. 3.4 Distribution of information on the identified EbA measures across the plan components (see text for further explanation)

where the gaps are. The figure shows that the 44 EbA measures identified in the plans can be grouped in six categories:

- Measures addressed in all the four plan components, from the *information base* through the *implementation*. This is obviously the most desirable situation, but it occurred only for 45.5% of the EbA measures. In all other cases, at least one component is lacking;
- Measures addressed in the first three components of the plans, but not in the *implementation* part. This occurs for 22.7% of the EbA measures;
- Measures addressed only in the *vision and objectives* and *actions* with no links to the *information base* or *implementation* (13.6%);
- Measures addressed only in the *information base* and *vision and objectives*, with no follow-up in the rest of the plan (6.8%);
- Measures addressed in the *information base* only, with no follow-up in the rest of the plan (2.3%)
- Measures addressed in the *vision and objectives*, *actions* and *implementation* components, with no links to the *information base* (2.3%).

3.4 Conclusions

The review concluded that maintaining/enhancing urban green spaces (e.g., ecological corridors, trees, gardens) is the most common measure, showing that there is strong awareness of the role that green areas play in addressing climate change challenges, both in terms of mitigating heat waves (measure *c*) and preventing floods (measure *f*).

The frequency of these measures is perhaps not surprising giving that they result in the enhancement of green areas, which is a typical objective that planners pursue to improve the urban space for a variety of purposes that go beyond climate change adaptation (e.g., providing recreation opportunities, improving air quality) (Tzoulas et al. 2007). So, their frequency could be explained by the fact that these measures rely on actions that are part of the standard approaches applied by planners for decades.

A general conclusion suggested by the review is that EbA measures are finding their way in climate adaptation plans, in response to a broad range of climate change challenges. However, a critical issue that we detected is that the proposal of these EbA measures in the plans is rarely backed-up by specific information on the expected outcomes, as well as the target beneficiaries. For example, the enhancement of green areas to reduce heat or to prevent floods is typically proposed as a general measure that will do some good, without providing details and justification for critical decisions, such as the design and the location of these interventions, and the distribution and vulnerability of the expected beneficiaries. Most plans are affected by a lack of specificity and details that may hamper the possibility for these measures to be actually implemented, as well as their overall effectiveness.

The baseline information upon which EbA measures are proposed and designed needs to be enhanced. Methods to assess the existing stock of green/blue infrastructures, and their potential to provide climate adaptation services must be mainstreamed in planning practice. Particularly, assessments of the flow of ES at local scales are often missing, given that many climate change impact and vulnerability studies provide results at larger scales, which limits their usefulness for developing local adaptation strategies (Vignola et al. 2009). A better knowledge base, including information on spatial pattern of vulnerability, would allow better targeting the design and implementation of EbA measures. The limited knowledge base used to design ES-related actions, as well as the lack of information about ES beneficiaries, have emerged as critical issues also in the review of urban plans presented in Chap. 2. The next two chapters address these issues. Chapter 4 illustrates a model that can help planners to assess the provision of a specific ES (micro-climate regulation), and to design urban green space accordingly. In Chap. 5, the outcomes of this and other ES models are combined with information on the potential beneficiaries to support urban planning interventions.

Chapter 4
Developing Ecosystem Service Models for Urban Planning: A Focus on Micro-Climate Regulation

Text and graphics of this chapter are based on: Zardo L, Geneletti D, Pérez-Soba M, Van Eupen M (2017) Estimating the cooling capacity of green infrastructures to support urban planning. Ecosyst Serv 26:225–235. https://doi.org/10.1016/j.ecoser.2017.06.016.

4.1 Introduction

Among the natural disasters occurring in Europe, heat waves cause the most human fatalities (EEA 2012). During the summer of 2003, for example, the heat wave in Central and Western Europe is estimated to have caused up to 70,000 excess deaths over a four-month period (EEA 2012). During the same period, in Germany alone, heat-related hospitalization costs had increased six-fold, not including the cost of ambulance treatment, while heat-related reduction of work performance caused an estimated output loss of almost 0.5% of the GDP (Hübler et al. 2008). In many regions of the world, climate change is expected to increase the effects of heat waves, including the rising of temperatures in cities (Koomen and Diogo 2015).

As shown in the previous chapter, the creation and enhancement of Urban Green Infrastructures (UGI) to regulate micro-climate and combat summer heat is one of the most common Ecosystem-based adaptation measure. By the virtue of their cooling capacity, i.e. capacity to modify temperature, humidity and wind fields, UGI can contribute to reducing high temperatures in cities, and lowering the related health risks (Lafortezza et al. 2013; Escobedo et al. 2015). Studies have shown that UGI have the capacity to mitigate high temperature in the summer, lowering them up to 6 °C (Souch and Souch 1993; McPherson et al. 1997). The creation and restoration of UGI, maximizing their cooling capacity, can reduce energy costs for air conditioning in summer and contribute to lowering mortality induced by higher temperatures (Koomen and Diogo 2015).

Urban plans represent a key governance instrument to design and enhance UGI (Kremer et al. 2013). However, as shown by our review in Chap. 3, despite the good awareness of the potential role of UGI to address climate change challenges, their inclusion in plans at the urban level often lacks sufficient baseline information (Geneletti and Zardo 2016). UGI may be very different in nature, including typologies such as parks, gardens, forests, green roofs and walls, and rivers (Naumann et al. 2011; Pauleit et al. 2011; EEA 2012). In turn, each typology may differ in key components (e.g. soil cover, tree canopy cover, size and shape), thus providing different ES, with different capacity (Bolund and Hunhammar 1999; Chang et al. 2007; de Groot et al. 2010; Bowler et al. 2010b).

This chapter presents an approach for estimating the cooling capacity provided by UGI tailored to support urban planning. The proposed approach, by providing guidance for UGI planning and design, is expected to support urban planners in effectively including the design and enhancement of UGI into the planning practice as a measure to cool cities and combat urban heat islands. In the remainder of the chapter, the approach is described, and applied to the city of Amsterdam, The Netherlands.

4.2   Methods to Assess the Cooling Capacity of UGI

As shown in Fig. 4.1, the proposed approach consists of five main steps. As a first step, the ecosystem functions of UGI that determine the cooling capacity are identified, following the cascade model (Haines-Young and Potschin 2010). Hence, the components associated to such ecosystem functions are defined, and their individual contribution to the cooling capacity assessed. Subsequently, the contributions are aggregate to determine the overall cooling capacity of the UGI. More specifically, the cooling capacity and the associated change in temperature are assessed for a set of UGI typologies, consisting of different combinations of tree canopy coverage, soil cover, and size. The proposed approach is based on an extensive analysis of the literature to determine the cooling capacity of UGI in three different climatic regions: Atlantic region, Continental region and Mediterranean region, as defined according to the classification of climate regions by the European Topic Center on Biological Diversity (ETC/BD 2006).

Shading, evapotranspiration (ETA) and wind shielding are the three ecosystem functions that determine the cooling capacity of UGI (EEA 2012; Gómez-Baggethun and Barton 2013; McPhearson et al. 2013; Smith et al. 2013; Larondelle and Haase 2013). In fact, vegetation regulates urban micro-climate in three ways: (i) by intercepting incoming solar radiation (shading), (ii) through the process of evapotranspiration, and (iii) by altering air movement and heat exchange. Shading and evapotranspiration are the ones that contribute the most to the cooling effect of UGI (Skelhorn et al. 2014). The contribution of wind to cooling capacity assessments, on the other hand, is rather complex to consider given its dependency on case-specific conditions (e.g. presence of buildings and directions of streets), that are not directly linked to ecosystem functions or components of UGI (Bowler et al. 2010b). Shading and evapotranspiration are determined by the structure of the ecosystem, i.e., the

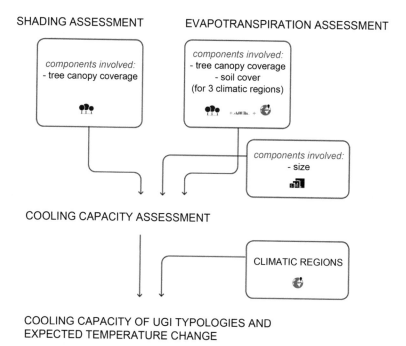

SHADING ASSESSMENT   EVAPOTRANSPIRATION ASSESSMENT

components involved:
- tree canopy coverage

components involved:
- tree canopy coverage
- soil cover
(for 3 climatic regions)

components involved:
- size

COOLING CAPACITY ASSESSMENT

CLIMATIC REGIONS

COOLING CAPACITY OF UGI TYPOLOGIES AND
EXPECTED TEMPERATURE CHANGE

Fig. 4.1 Flowchart of the proposed approach to assess the cooling capacity of UGI

architecture of its components. In particular, as shown in Fig. 4.1, shading and evapotranspiration are associated to three specific components of UGI; namely, *tree canopy coverage*, *soil cover*, and *size* (Taha et al. 1991; Akbari et al. 1992; Souch and Souch 1993; Chang et al. 2007; Cao et al. 2010; Bowler et al. 2010b; Schwarz et al. 2011; Larondelle and Haase 2013).

4.2.1 Shading and Evapotranspiration Assessment

Several studies show evidence of cooler air temperature beneath individual or clusters of trees, highlighting the amount of shading as an important factor affecting temperatures (Taha et al. 1991; Akbari et al. 1992; Bowler et al. 2010b; Schwarz et al. 2011; Larondelle and Haase 2013). Among the indicators proposed in the literature, the *tree canopy coverage*, expressed as the percentage of the ground area shaded by tree canopies relative to the total open area, is here adopted (Potchter et al. 2006; Strohbach and Haase 2012). Accordingly, assuming a linear relationship between the presence of tree covers and shading (Potchter et al. 2006), the assessment of the contribution of shading can be based on a visual estimation. For example, a shading score equal to "x" is assigned to an UGI with an x% tree canopy coverage. Noteworthy, the contribution given by trees with canopy lower than two meters is here overlooked, as they do not provide shade that is useful for human beneficiaries. Nevertheless, such vegetation has a significant contribution in terms of evapotranspiration.

Tree canopy coverage, soil cover and tree species are the three main components that jointly affect evapotranspiration (Taha et al. 1991; Akbari et al. 1992; Souch and Souch 1993; Allen et al. 1998; Bowler et al. 2010b; Schwarz et al. 2011; Larondelle and Haase 2013). In this context, the focus is however on the first two components mainly because information of tree species is hardly available at city scale, whereas, at this scale, differences in evapotranspiration across different combinations of species can be considered negligible (Souch and Souch 1993). On the other hand, the *climatic region* is included as a crucial factor to consider given that it greatly affects the evapotranspiration: in warm and dry areas, evapotranspiration is more effective than in humid or cool climates (Taha et al. 1991; Akbari et al. 1992; McPherson et al. 1997; Bowler et al. 2010b).

Following the approach by Allen et al. (1998), the evapotranspiration is calculated as:

$$ET_c = K_c \bullet ET_0$$

where ET_c is the tree or soil cover evapotranspiration (ETA) under conditions of unlimited presence of water in the ground (irrigated), K_c is the tree or soil cover coefficient, and ET_0 is the reference evapotranspiration, which takes into account the climatic region of the study area.

Operationally, to estimate the evapotranspiration potential of an UGI, its soil cover and tree canopy coverage are analysed separately to determine the related Kc coefficient. Hence, the evapotranspiration is estimated by multiplying the Kc coefficient by the climate-specific value ET_0, again analysing separately the contribution of trees and of the soil cover (e.g., Kremer et al. 2013; Larondelle and Haase 2013; Schwarz et al. 2011). In the proposed approach, the overall evapotranspiration value of the UGI, obtained by adding the different contribution that are expressed in mm d^{-1}, is then standardized into an evapotranspiration score in the 0–100 range.

4.2.2 Cooling Capacity Assessment

The extent to which shading and evapotranspiration contribute to the overall cooling capacity of the UGI is determined by the size of the UGI itself (Chang et al. 2007; Cao et al. 2010; Bowler et al. 2010b). In fact, shading and evapotranspiration jointly reduce the air temperature, but the impact of evapotranspiration becomes predominant as the area gets larger (Akbari et al. 1992). Specifically, green areas larger than 2–3 ha are cooler than their surroundings, whereas green areas smaller than 2 ha have a limited effect (Chang et al. 2007). Thus, several studies identify the threshold between *small* parks and *large* parks around a value of 2 ha (e.g., Bowler et al. 2010b; Cao et al. 2010; Chang et al. 2007; Shashua-Bar and Hoffman 2000).

Regarding the effects of shading and evapotranspiration on surroundings of trees (based on measurements taken at 12 m and 5 m from trees), Akbari et al. (1992) concluded that, for large areas, the cooling capacity depends mainly on evapotranspiration, reaching a distance as far as five times the height of the tree. They also found that shading contributes up to 95% when directly under the canopy, but its contribution in terms of reducing the temperature (and consequently the energy consumption for air conditioning) is around 40% for areas larger than 2 ha. According to Chang et al. (2007), size contributes to 60% of the cooling capacity, and indirectly affects the contribution of ETA. Finally, Shashua-Bar and Hoffman (2000), based on empirical studies, note that in areas smaller than two hectares, the contribution of shading is around 80% of the total cooling capacity, with the remaining 20% determined by evapotranspiration.

Therefore, in the proposed approach, the overall cooling capacity of UGI is assessed through a weighted summation of the evapotranspiration and shading scores, using different weights according to size, followed by a standardization of the results into a scale between 0 and 100. More specifically, in areas smaller than two hectares, shading is assigned a weight of 0.8 and evapotranspiration of 0.2, while in areas larger than two hectares, the weights are of 0.4 and 0.6, for shading and ETA, respectively. Noteworthy is the case of areas with less than 50% of tree canopy coverage that may turn to be warm islands instead of cool islands during some part of the day in very hot summer (Chang et al. 2007). To consider this remark, the cooling capacity scores calculated for all areas with tree canopy coverage below 50% is marked with a "*" to highlight that, in some circumstances, they can also work the other way round.

4.2.3 UGI Typologies and Expected Temperature Change

To define different typologies of UGI, the three components of tree canopy coverage, soil cover, and size, were combined. To this end, tree canopy coverage is classified into five intervals: 0–20%, 21–40%, 41–60%, 61–80% and 81–100%. Soil cover is classified into sealed (all impervious surfaces), bare soil, heterogeneous cover (mixed cover of bare-soil and shrubs, typical of vegetable gardens or inner courts or some vacant lots), grass (fine vegetation), and water, based on the HERCULES soil-cover taxonomy (Cadenasso et al. 2007). Finally, size was divided into two classes: below and above two hectares. By combining these classes of the three components, 50 typologies of UGI are obtained to be further analysed considering the three different climatic regions.

Operationally, to assess the cooling capacity of each UGI typology in each climatic region, data on ET0 and Kc was retrieved from a number of databases, including the CGMS database of the Mars Crop Yield Forecasting System and the FAO (more details in Zardo et al. 2017). Through a literature review, the cooling capacity score of each UGI typology was then associated to an expected change in temperature (see Table 4.1). Indeed, the conversion of cooling capacity scores (from 0 to 100)

Table 4.1 Overview of some key references about cooling capacity of UGI

Climatic area	Min cooling (°C)	Max cooling (°C)	Reference
Atlantic (Koppen: Cfb)	1.0	3.5	Watkins (2002); GLA (2006); Schwarz et al. (2011); Larondelle and Haase (2013)
Continental (Koppen: Cfa)	1.0	4.8	Potchter et al. (2006); Chang et al. (2007)
Mediterranean (Koppen: Csa)	1.7	6.0	Taha et al. (1991); Souch and Souch (1993); Shashua-Bar and Hoffman (2000); Potchter et al. (2006)

into changes in air temperature is significantly affected by the climatic region: UGI can lower daily maximum near-surface temperature especially in hot and dry conditions (Taha et al. 1991).

4.3 Assessing the Cooling Capacity of UGI Typologies

Figure 4.2 summarizes the assessment of the cooling capacity of 50 different typologies of UGI, for the three climatic regions. From the analysis of the table it is possible to note that 26% of the UGI typologies have the highest scores (from 81 to 100), 17% of UGI typologies score between 61 and 80, 23% from 41 to 60, 23% from 21 to 40, and 12% score from 0 to 20. Indeed, size is the most influent component among the three; for example, all UGI scoring more than 60 have a size above the two hectares, while only 3% of the UGI that are above the two hectares score less than 60 below. Furthermore, no UGI with size smaller than two hectares scores more than 60. While scores between 41 and 60 comprise mainly areas smaller than two hectares (91%). This includes all the UGI with 100% of tree canopy coverage, and most of UGI with 80% of tree canopy coverage. The UGI scoring between 21 and 40 are smaller than two hectares and have tree canopy coverage between 20 and 60%. Thus, the second most influential component, after the size, is the tree canopy coverage, followed by soil cover.

In terms of the expected temperature changes, according to the climatic region, each score implies a different temperature decrease. As shown in Fig. 4.3, the Mediterranean region is where larger changes occur, followed by the Continental area and the Atlantic region. For example, an UGI with a cooling capacity score of 100, in the Atlantic region, can decrease the temperature up to 3.5 °C, while the same UGI, in the Mediterranean region, can reduce the temperature of up to 6 °C. Therefore, investing on UGI to improve their cooling capacity has different implications in terms of temperature decrease, depending on the climatic region (e.g., see grey rectangle in Fig. 4.3).

tree canopy coverage	soil cover	ATLANTIC REGION size: < 2 ha	> 2 ha	CONTINENTAL REGION size: < 2 ha	> 2 ha	MEDITERRANEAN REGION size: < 2 ha	> 2 ha
20%	sealed	11	20	11	20	11	20
	bare soil	18	65	17	58	17	56
	heterogeneous	19	68	19	68	19	66
	grass	19	70	19	69	20	74
	water	20	75	20	75	20	72
40%	sealed	22	40	22	40	22	39
	bare soil	27	74	27	69	26	67
	heterogeneous	28	76	28	76	27	74
	grass	28	78	28	66	28	80
	water	28	81	29	81	28	79
60%	sealed	29	60	33	60	33	59
	bare soil	33	83	36	79	36	77
	heterogeneous	36	84	37	84	36	82
	grass	37	85	37	84	37	86
	water	37	87	37	87	37	85
80%	sealed	37	80	44	80	43	78
	bare soil	44	91	45	90	45	88
	heterogeneous	46	92	46	92	45	90
	grass	46	93	46	92	46	92
	water	46	94	46	94	46	91
100%	sealed	55	100	55	100	54	98
	bare soil	55	100	55	100	54	98
	heterogeneous	55	100	55	100	54	98
	grass	55	100	55	100	54	98
	water	55	100	55	100	54	98

legend:
0-20
21-40
41-60
61-80
81-100

Fig. 4.2   Cooling capacity estimated for 50 UGI typologies in the climatic regions

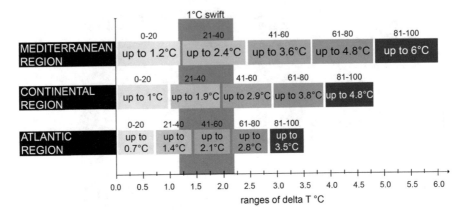

Fig. 4.3 Temperature variation (in Celsius degrees) for the same classes in the different climatic regions

4.4   Application to the City of Amsterdam

By way of example, the proposed approach was applied to the city of Amsterdam. A 10 × 10-km study area was selected to analyse the existing UGIs and to assess their cooling capacity. Amsterdam belongs to the cold temperate moist zone, corresponding to the Atlantic climatic region. Tree canopy coverage, soil cover and UGI size were mapped using available data (Fig. 4.4).

Overall, 74,653 patches, covering 8477 hectares, were mapped. The mapped UGI consist almost entirely (90%) of water and of green areas with a tree canopy coverage below 20%. A heterogeneous soil cover with tree canopy coverage below 20%, sealed patches with tree canopy coverage below the 20%, sealed patches with tree canopy coverage between 20 and 40%, and grass soil cover with tree canopy coverage below 20% characterize the remaining 10% of the mapped UGI. Accordingly, Fig. 4.5 presents the results of the cooling capacity assessment with scores from 0 to 100. Most UGI have a low cooling capacity: 34% of total UGI score less than 25, 13% score between 25 and 30, 22% score between 30 and 60, and only the 1% of the UGI score more than 60. As for the potential temperature reduction, it can be assumed that the 22% of UGI, which score between 60 and 30, can lower the temperature up to 2.1 °C.

From an urban planning perspective, the results can be used to identify possible actions to enhance the cooling capacity of least performing UGI. By way of example, Fig. 4.6 presents a set of possible interventions to upgrade an UGI with a cooling capacity score of 11, characterized by size below two hectares, 20% of tree canopy coverage and sealed soil. The best results are provided by a combination of actions targeted at increasing the size and the tree canopy coverage, and improving soil cover.

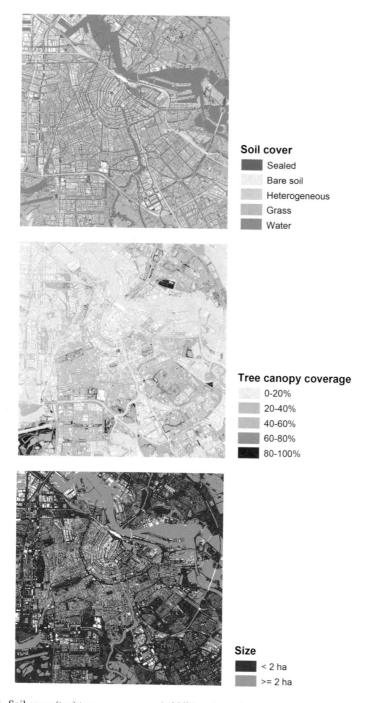

Soil cover
◼ Sealed
▫ Bare soil
▨ Heterogeneous
▨ Grass
◼ Water

Tree canopy coverage
▫ 0-20%
▨ 20-40%
▨ 40-60%
▨ 60-80%
◼ 80-100%

Size
◼ < 2 ha
▨ >= 2 ha

Fig. 4.4 Soil cover (top) tree canopy cover (middle) and size (bottom) of UGI in Amsterdam

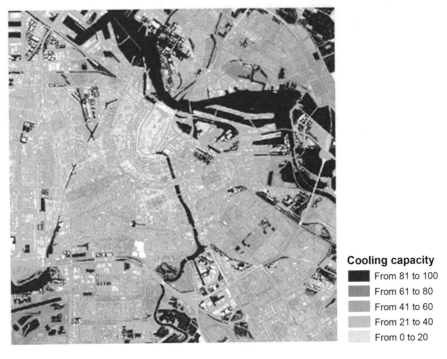

Cooling capacity
- From 81 to 100
- From 61 to 80
- From 41 to 60
- From 21 to 40
- From 0 to 20

Fig. 4.5 Map showing the cooling capacity of UGI in Amsterdam

4.5 Lessons Learned and Conclusions

Scientific knowledge from different fields, such as ecology, planning, urban forestry and climate science, can improve UGI assessment, but an effort in terms of converting it into guidance that can improve urban planning processes is still needed (Norton et al. 2015). This chapter showed an example of how to draw on knowledge and data from different disciplines to improve the understanding of the relationship between the characteristics of UGI and the provision of ES. UGI represent a great potential for cities to adapt to multiple challenges; hence, the importance to take into account the design of UGI, and their capacity to supply a range of ES, in urban planning (Munang et al. 2013). However, the lack of suitable data and the complexity of modelling tools often pose a challenge to the improvement of UGI design in planning exercises. In this chapter, out of the bundle of ES provided by UGI, we focused on micro-climate regulation. The proposed approach links the relevant ecosystem functions and components of a UGI to its capacity to reduce heat, distinguishing among different typologies of UGI. The approach is designed to fit the urban scale and to work with input data that are sufficient to differentiate among the cooling capacity provided by different types of UGI, but still easily available during urban planning processes.

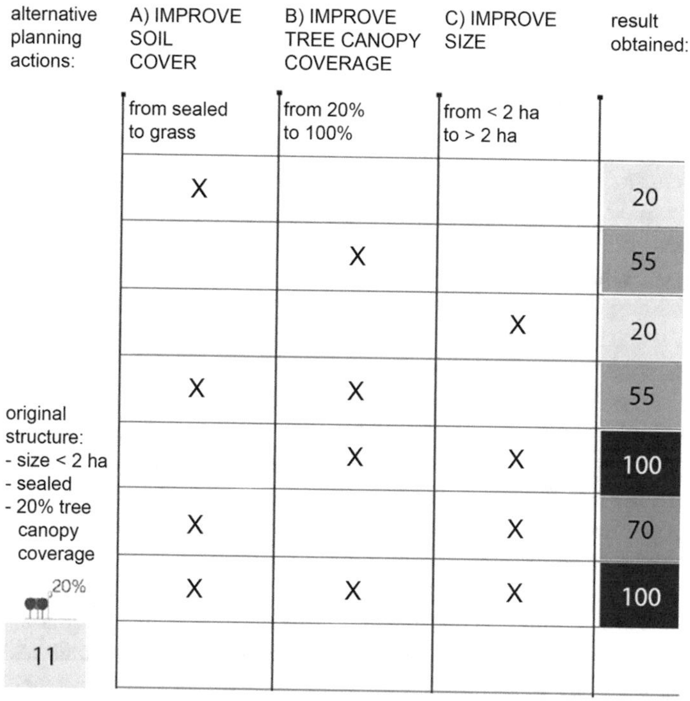

alternative planning actions:	A) IMPROVE SOIL COVER	B) IMPROVE TREE CANOPY COVERAGE	C) IMPROVE SIZE	result obtained:
	from sealed to grass	from 20% to 100%	from < 2 ha to > 2 ha	
	X			20
		X		55
			X	20
original structure:	X	X		55
- size < 2 ha		X	X	100
- sealed	X		X	70
- 20% tree canopy coverage	X	X	X	100
20% 11				

Fig. 4.6 Alternative actions (described in columns A, B and C) to upgrade an hypothetical UGI with a cooling capacity scores below 20 (characterized by a size smaller than two hectares, sealed soil cover and tree canopy coverage of 20%), and the expected improvement in terms of cooling capacity score (last column)

As described in the chapter, the three most relevant components of UGI (i.e. tree canopy coverage, soil cover, and size) do not equally determine the cooling capacity. Generally, the most important component is size, followed by tree canopy coverage and lastly soil cover. Similarly, the climatic region of the study area is very important in determining the decrease in air temperature (°C) provided by UGI. In particular, it was noted how a given UGI, with a specific class of cooling capacity, implies a different air temperature reduction depending on the climatic region. In the Mediterranean region, the same UGI can lower the temperature more effectively than in Atlantic or Continental regions, with the consequent different investment implications from a practical point of view.

The application to the city of Amsterdam demonstrated that the proposed approach requires only a limited set of input data, generally easy to obtain, to provide an overall cooling capacity assessment of the UGI. Several practical insights emerge from the case study application that are related to the different effects of the components in different conditions. For example, shading is more important in small areas than in large areas, making the increase of tree canopy coverage particularly interesting for small green spaces, especially compared to soil cover interventions. The latter are more adequate for large areas, while for small areas an enhancement of the cooling capacity can be obtained by increasing the tree canopy cover. An exception is the Mediterranean region where trees are more preferable than soil cover interventions. On the contrary, for large areas soil cover changes can provide much more interesting results in all three climatic regions, especially in the Mediterranean. However, a good balance in terms of tree-canopy coverage, soil cover type and size, as mentioned in the previous paragraph, is the strategy providing the best cooling capacity.

The approach, as it stands now, has three main limitations. Firstly, the computation of shading, evapotranspiration and the overall cooling capacity is based on a review of the available literature and on expert opinion. Further empirical evidence would make the approach stronger. Secondly, variables such as wind flow, city morphology, and tree species were not considered due to the choice for simplicity and synthesis, looking for a fair trade-off between accuracy of the assessment and a complexity in computations and data. Despite restricted to the most influencing factors, the analysis is flexible enough to provide solutions that are site-specific. For example, concerning tree species, the literature provides evidence about the fact that different tree species differently contribute to cooling due to different evapotranspiration functioning. Last, the proposed approach only considers the cooling capacity within the UGI, without addressing the effects outside its boundaries. Clearly, knowing the spatial extent of the cooling capacity beyond UGI boundaries would be interesting for urban planning, and for an analysis of the expected beneficiaries of different interventions. This challenge is addressed in the next chapter, where the analysis of the cooling capacity of UGI, as well as of other ES, is linked to an explicit assessment of different groups of beneficiaries, and the outcomes are used to inform urban planning.

Acknowledgments   Marta Pérez-Soba and Michiel Van Eupen are acknowledged for contributing to this chapter.

Chapter 5
Applying Ecosystem Services to Support Planning Decisions: A Case Study

Text and graphics of this chapter are based on: Cortinovis C, Geneletti D (2018) Mapping and assessing ecosystem services to support urban planning: A case study on brownfield regeneration. One Ecosyst 3:e25477. doi: https://doi. org/10.3897/oneeco.3.e25477

5.1 Introduction

Although several authors acknowledge the potential of ecosystem service (ES) assessments to increase the quality of planning processes and decisions (Geneletti 2011; Mckenzie et al. 2014; Rall et al. 2015), most urban ES studies are still far from real-life application. While urban ES research demonstrates continuous methodological advancements, scientific works often lack the identification of specific policy questions and stakeholders to which they might be relevant (Haase et al. 2014), thus resulting in generic and abstract recommendations with no direct applicability to the planning and management of green infrastructure (Luederitz et al. 2015). The aim of this chapter is to show how ES knowledge, i.e. information produced by ES assessments (see Chap. 4), can be used to support decision-making in a real-life urban planning context.

ES knowledge can enter policy- and decision-making processes through multiple pathways associated to different potential impacts, from raising stakeholder awareness to shaping specific decisions. (Posner et al. 2016). Among the pathways described by Posner and colleagues, we refer here to the use of ES knowledge to *generate actions* and *produce outcomes*. The expected result is the establishment of new or updated plans and policies that consider impacts on ES and promote their balanced provision, ultimately improving human health and wellbeing along with biodiversity and nature conservation (Posner et al. 2016).

Drawing from a set of case studies, Barton et al. (2018) provide some examples of the tasks that ES assessments can perform in these contexts. ES knowledge can assume a *decisive* role, when it is used to support the formulation and structuring of the decision problem; to identify criteria for screening, ranking, and spatial-targeting of the alternatives; or to provide arguments for negotiations, shared norms, and conflict resolution. It can also assume a *design* role, when it is used to set the basis for implementation tools, including the definition of standards and policy targets; the design of regulations, certifications, pricing, and incentives; or the establishment of damage compensations (Barton et al. 2018).

The *decisive* role encompasses the use of ES knowledge in the specific phase of urban planning processes when alternative scenarios must be assessed and compared (e.g., Kain et al. 2016). This use poses specific requirements to ES assessments. First, it entails identifying appropriate indicators that measure the expected outcomes of planning actions in terms of changes in human wellbeing, coherently with planning objectives (Ruckelshaus et al. 2015). Second, it requires assessing the consequences of planning interventions on multiple ES, explicitly addressing the potential trade-offs between different ES and competing land uses that characterize the alternative planning scenarios (Sanon et al. 2012; Woodruff and BenDor 2016; Kain et al. 2016).

This chapter presents an application in which ES assessments are used to support a real-life planning decision about the regeneration of brownfields in the city of Trento (Italy). Potential re-greening interventions are prioritized according to their expected consequences on two illustrative, though relevant ES for the city. The performance of the different alternatives in relation to the two ES is assessed by comparing current conditions and future planning scenarios. Then, the results are combined through a multi-criteria analysis where criteria correspond to a set of defined planning objectives that may assume different weights according to different stakeholder perspectives.

5.2   Case Study: Brownfield Regeneration in Trento

Trento is an alpine city of around 120,000 inhabitants in northeastern Italy, located along the valley of the river Adige, roughly half-way between the Brenner Pass and the Adriatic Sea. The main settlement originated from the concentration of urban areas and infrastructures in the valley floor and hosts around 70% of the population. The remaining 30% lives in small villages spread across the surrounding hills and mountains. Agricultural areas, predominantly vineyards and apple orchards, occupy the few non-urbanized patches on the valley floor and the sunny hillsides. Forests cover almost half of the large municipal territory, which spreads over more than 150 km^2 and up to an elevation of 2180 m. Of this, more than 10 km^2 are designated as natural protected areas, including eight Natura 2000 sites and four local reserves (Fig. 5.1).

The presence of brownfields is one of the main planning issues in Trento. The current urban plan identifies thirteen 'urban redevelopment areas', mostly former industrial sites or partially-abandoned residential blocks, ranging in size from 0.5 to 9.9 ha and covering a total area of around 44 ha. With few exceptions, they are close to the most dense and populated parts of the city, where their presence exacerbates

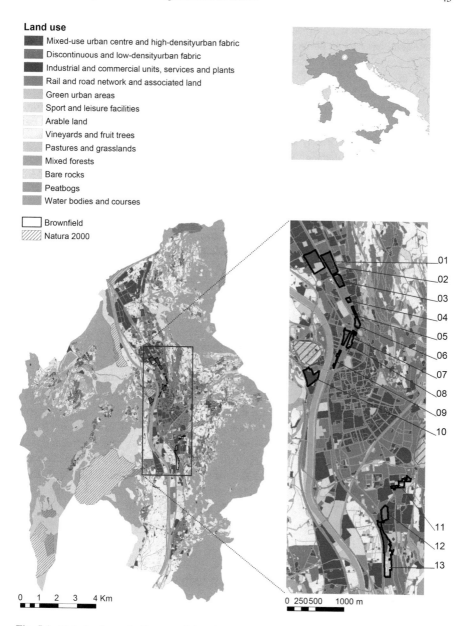

Fig. 5.1 Main land uses in Trento and the 13 brownfields identified by the urban plan as 'urban redevelopment areas'

social, economic, and environmental problems, but at the same time is an opportunity to benefit through regeneration interventions a large part of the population. Until now, the costs, especially for contaminated sites, and the bureaucratic burden associated to intervention, as well as the sometimes-contrasting interests of public administration and private owners, have hindered their transformation.

Considering the existing situation, it is reasonable to assume that only some of the brownfields will be converted to new industrial or residential areas in the next years. A greening intervention can thus be advanced as a possible, perhaps temporary, solution. Accordingly, the study hypothesizes a possible conversion of the thirteen urban redevelopment areas into new public parks, with the aim of identifying which intervention should be prioritized to maximise the benefits for the surrounding population. Benefits are measured in terms of increased provision of two key urban ES for Trento, namely micro-climate regulation and nature-based recreation.

The selection of micro-climate regulation responds to the growing concerns for summer heat waves, particularly intense in the city due to its low altitude and to the narrowness of the valley. During the 2003 event, Trento proved to be more vulnerable to heat waves than other Italian cities (Conti et al. 2005). The combined effect of heat waves and of the intense urban heat island in the most urbanized part of the city causes peaks in energy demand and threatens citizens' health and wellbeing (Giovannini et al. 2011). Considering the increased frequency and intensity of heat waves expected in the coming decades (Fischer and Schär 2010), effective solutions to control the urban micro-climate and to provide cool areas for heat relief during the hot season are seen as one of the most pressing needs by citizens and administration.

The selection of recreation responds to the specific planning goal of the city administration: to provide equal opportunities for nature-based recreation and relaxation to all citizens. During the last years, new public parks have been realized in peri-urban areas to gain a more balanced distribution over the city. However, understanding if opportunities for nature-based recreation are equally distributed is not an easy task. In Trento, besides urban parks, citizens also benefit from the proximity to other typologies of green areas where they conduct a wide range of day-to-day recreational activities, including hiking, mountain-biking, trail running, and climbing. Indicators based on the availability of and accessibility to public urban parks, though common in urban planning applications, are not enough to capture this variety and to support planning decisions (Cortinovis et al. 2018). Assessing recreation as an ES, considering different providing units and different levels of demand, could provide planners with useful information for achieving an equal distribution of recreational opportunities over the city (Kabisch and Haase 2014).

5.3   Producing ES Knowledge to Evaluate Planning Scenarios

5.3.1   Mapping and Assessing ES

The cooling effect of urban green infrastructure was assessed by applying the method described in Chap. 4 (see also Zardo et al. 2017 for furher details), specifically designed to support planning and management decisions at the urban and suburban scale. Green infrastructure in Trento were classified according to their size, soil cover, and percentage of canopy coverage, and assigned a cooling capacity score and the respective class based on Fig. 4.2. Then, the cooling effect produced on the surroundings was mapped by approximating the effect of evapotranspiration through linear decay functions that vary depending on the size of the area, and the effect of shading through local buffers around canopies (Geneletti et al. 2016). The final map of the cooling effect is divided into six classes, from A+ (maximum cooling effect) to E (minimum or no cooling effect). The difference between two successive classes corresponds approximately to 1 °C.

Opportunities for nature-based recreation in the city were assessed through a locally-adjusted version of ESTIMAP-recreation, a model originally developed for mapping ES across Europe (Zulian et al. 2013; Paracchini et al. 2014) and later adjusted for the application to different contexts and scales (Baró et al. 2016; Liquete et al. 2016; Vallecillo et al. 2018; Zulian et al. 2018). The model is composed of two modules. The first module assesses the Recreation Potential (RP), i.e. the suitability of different areas to support nature-based recreational activities based on their intrinsic characteristics. Elements that contribute to define the RP are identified in the three categories of natural features, urban green infrastructure components, and land uses. The map of RP is a raster with values ranging from 0 (no RP) to 1 (maximum RP in the analysed area). The second module assesses the Recreation Opportunity Spectrum (ROS), i.e., the actual opportunities for nature-based recreation offered to the citizens. The value of ROS is obtained by combining RP with information about proximity, here defined as the availability of infrastructures and facilities to access (e.g., cycle paths, bus routes, parking areas) and to use (e.g. playgrounds, sport fields, park furniture) the areas. The map of ROS is classified into nine categories resulting from the cross-tabulation of high/medium/low RP and high/medium/low availability of infrastructure and facilities.

To produce the maps of RP and ROS, the elements considered in each module are spatially combined according to scores assigned by the user. In the described application, scores were elicited from a pool of seventeen local experts, including key personnel of different provincial and municipal departments, researchers from different institutions, and local practitioners. The experts were asked to fill-in an online questionnaire and then invited to discuss the preliminary results in a follow-up focus group. A detailed list of the data used for the analysis and the description of the involvement process, including the final scores used to run the model, can be found in Cortinovis et al. (2018).

5.3.2 Comparing Scenarios Based on ES Beneficiaries

The results of the analysis at the city scale were used as a baseline to assess the potential benefits produced by brownfield regeneration. The conversion of each of the thirteen brownfields into a new green area was considered as an alternative planning scenario and analysed independently. The outcome of the transformation was assumed to be, for each brownfield, a new urban park, intensely planted and open to public use. Accordingly, to assess their expected cooling effect, new urban parks were modelled as areas covered by grass with canopy coverage ranging from 80%

to 100%. To assess opportunities for nature-based recreation, brownfields were assigned to the land use class 'green urban areas', assuming the same presence of infrastructures and facilities as in other parks of comparable size.

Similar indicators, based on the number of people affected by the transformation, were used to assess the two ES. For each ES, vulnerable people, defined as citizens' groups with a higher-than-average need for the specific service, were identified and quantified as a sub-group of the total beneficiaries. In the case of cooling, beneficiaries were defined as citizens that experienced a positive change in the class of cooling effect of their living place. Young children (< 5 years old) and the elderly (> 65 year old) were selected as vulnerable groups, based on their higher sensitivity to heat stress (Basu and Samet 2002; Kenny et al. 2010; Kabisch et al. 2017). In the case of recreation, people living within 300 m from the new parks were considered as beneficiaries (Kabisch et al. 2016; Stessens et al. 2017). Children and teenagers (< 20 years old) and the elderly (> 65 year old) were identified as vulnerable groups, based on the higher demand for close-to-home recreation and relaxation areas (Kabisch and Haase 2014). Furthermore, those beneficiaries already served by high-level opportunities for nature- based recreation in the current condition (i.e., living within 300 m from areas classified in the highest class of ROS), were counted separately.

The results of the two ES assessments were combined through a multi-criteria analysis, using the thirteen scenarios as alternatives and the two ES and corresponding categories of beneficiaries as criteria and sub-criteria, respectively (Table 5.1). Weights were assigned according to three illustrative policy perspectives and related objectives. The 'cool air for the elderly' perspective prioritizes the improvement of the cooling effect in areas with a high share of older population. The 'every child needs a park' perspective favours opportunities for recreation to people, especially

Table 5.1 The three policy perspectives and respective combinations of weights considered in the multi-criteria analysis to prioritize brownfield regeneration scenarios

	Perspective 1 "balanced"			Perspective 2 "cool air for the elderly"			Perspective 3 "every child needs a park"		
Cooling	0.50			0.80			0.20		
Non-vulnerable		0.20			0.14			0.20	
< 5 years old		0.40			0.29			0.40	
> 65 years old		0.40			0.57			0.40	
Recreation	0.50			0.20			0.80		
Non-vulnerable		0.20			0.20			0.14	
Served			—			—			0.20
Not served			—			—			0.80
< 20 years old		0.40			0.40			0.57	
Served			—			—			0.20
Not served			—			—			0.80
> 65 years old		0.40			0.40			0.29	
Served			—			—			0.20
Not served			—			—			0.80

children and teenagers, who are not served in the present condition. The 'balanced' perspective promotes both ES equally but assigns a higher weight to vulnerable groups. Values for each criterion and sub- criterion were normalised according to the maximum and a 'weighted summation' approach was used to calculate the over- all score for each alternative, hence defining the final rankings for the three perspec- tives. Finally, a sensitivity analysis was conducted to explore the robustness of the rankings to variations in the weights assigned to criteria and sub-criteria. Further details on the methodology adopted to compare the brownfield regeneration sce- narios can be found in Cortinovis and Geneletti (2018a).

5.4 Planning Scenarios Evaluated Through ES

5.4.1 Current ES Provision Across the City

The current provision of the two analysed ES (Figs. 5.2-5.3) is used as baseline to measure the benefits of brownfield regeneration scenarios. The map of the cooling effect produced by green infrastructure in the valley floor, i.e., the most urbanised area of the city, shows a prevalence of the highest classes of cooling effect (Fig. 5.2). The presence of close-by forests and of the river Adige and its tributaries contributes to lower the temperature of their surroundings. Disadvantaged areas can be observed in the dense neighbourhoods close to the city centre and in the northern suburbs. Here, the mix of residential and industrial areas with little green infrastructure, as well as the high rate of soil sealing produced by the concentration of urban activities and major transport infrastructures, have a negative impact on the cooling perfor- mance of the city. Most of the brownfields are strategically located close to areas that scarcely benefit from the cooling effect of green infrastructure.

The map of ROS shows a clear difference in the recreation opportunities between the main urban settlement in the valley floor and its surroundings (Fig. 5.3). The main urban settlement is mostly characterized by low values of RP, but large urban parks and the riverbanks provide good opportunities for nature-based recreation thanks to their high accessibility. The surroundings are mostly characterised by high values of RP, but the availability of infrastructures and facilities is not homoge- neous. Forests characterised by the presence of forest tracks, hiking trails, and facilities dedicated to specific activities such as climbing routes and MTB trails are mainly located close to the villages. Areas in the highest class of ROS prevail on the east side of the valley, whilst on the west side, where the settlements are sparser and the connections with the valley floor are more difficult, many areas are characterised by high RP but low proximity. The analysis suggests different directions for inter- ventions. While infrastructures and facilities could be strengthened in natural and semi-natural areas, interventions in the main urban settlement should focus on achieving a more equal distribution of urban green areas. The location of some of the brownfields represents an opportunity to act in this direction, thus enhancing the condition of people that currently have no or very few close-to-home opportunities for nature-based recreation.

Cooling class

- A+
- A
- B
- C
- D
- E

Brownfield

0   1   2   3   4 Km

Fig. 5.2 Map of the cooling effect of urban green infrastructure in the most urbanized part of the city of Trento

5.4.2 Potential Benefits of Brownfield Regeneration

An example of how the conversion of brownfields into new urban parks would affect the two analysed ES is provided in Fig. 5.4. Due to the change in the soil cover from partly sealed to grass and to the intense plantation, the site would reach the highest class of cooling effect, thus also positively affecting the micro-climate of the immediate areas. While in the present condition, most of the

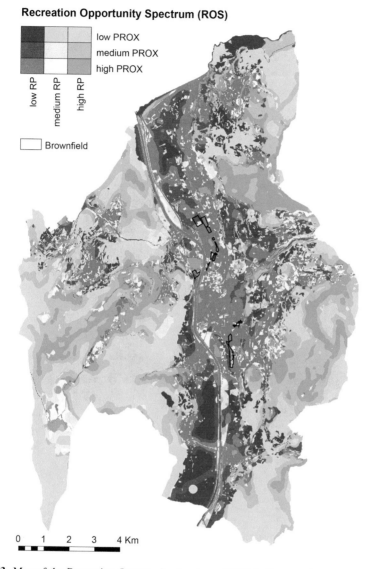

Recreation Opportunity Spectrum (ROS)

Fig. 5.3 Map of the Recreation Opportunity Spectrum (ROS) in Trento calculated through the locally- adjusted version of the ESTIMAP-recreation model

surrounding residents gain very little or no thermal benefit at all from the presence of green infrastructure, almost exclusively limited to single shading trees, in the regeneration scenario the major part of the area is affected by a noticeable improvement. In the neighbourhood to the North, some households would shift from the lowest to the highest class of cooling effect. The conversion to a new urban park would also be an opportunity to increase the availability of public green areas by connecting the converted brownfield to the adjacent open-air

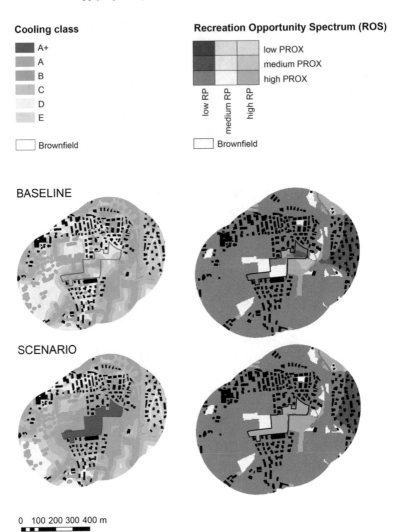

Fig. 5.4 An example of comparison between baseline condition and planning scenario for brownfield 11. Maps of the cooling effect considering the maximum distance potentially reached by the cooling effect produced by the brownfield (left) and maps of the Recreation Opportunity Spectrum showing the 300-m buffer used to identify potential beneficiaries (right)

soccer field. In the regeneration scenario, all the households included in the map would benefit from an additional space for recreation within walking distance from their location.

The different scenarios are assessed and compared based on the number of people that would benefit from brownfield regeneration (Fig. 5.5). In terms of cooling effect, scenario 11, involving the conversion of a large brownfield inside a densely

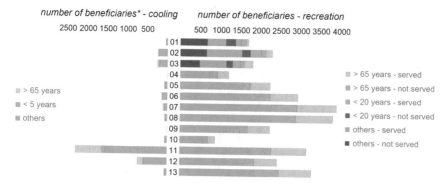

Fig. 5.5 Expected benefits produced by the different scenarios in terms of enhanced cooling effect by urban green infrastructure (left) and enhanced opportunities for nature-based recreation (right): number of beneficiaries broken down into different vulnerability classes

built-up and populated part of the city, is by far the best performing one: more than 2000 citizens would benefit from the enhanced cooling effect produced by the new green area. The other scenarios are expected to affect from some decades to few hundreds of people. Overall, the number of people that would benefit from increased opportunities for nature-based recreation is much higher. Scenarios 07 and 08 produce the highest absolute number of beneficiaries. However, only the regeneration of brownfields 01, 02 and 03 would serve people that, at present, have no access to close-to-home nature-based recreational opportunities. The ratio between total beneficiaries and specific vulnerable groups is not the same across scenarios, due to the uneven distribution of population groups across the city. For example, the share of children and teenagers is higher for scenarios 01 and 02 compared to the others, while the share of people aged more than 65 is the highest for scenario 11.

The information about the number of beneficiaries of the two ES in the different scenarios was combined through a multi-criteria analysis according to the three perspectives described in Table 5.1 (Fig. 5.6). When assuming a 'balanced' perspective, with the same weight assigned to the two ES and a double weight assigned to vulnerable compared to non-vulnerable groups, scenario 11 ranks first. The second perspective, consistent with the objective of improving the cooling effect in areas with a high share of older population, leads to the same first-ranking scenario. Although the other positions change between the two perspectives, all scenarios gain a very low score compared to scenario 11. Under the third perspective, the final ranking changes significantly and the first positions are occupied by the three scenarios (01, 02 and 03) involving the regeneration of brownfields located in the northern part of the city, where the population is comparatively younger and existing opportunities for recreation are scarcer.

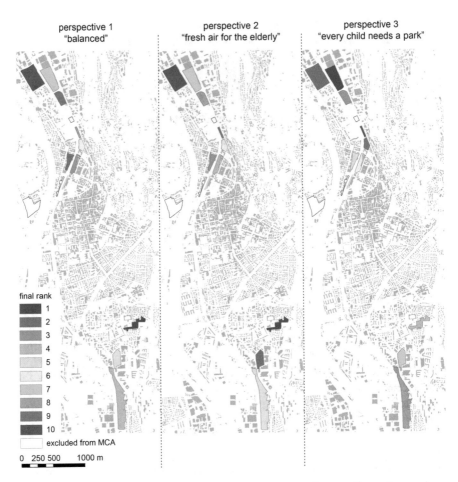

Fig. 5.6 Map of the priority level of brownfield regeneration scenarios according to three perspectives considered in the multi-criteria analysis

5.5   Lesson Learned and Conclusions

The case study showed a possible use of ES knowledge to support urban planning in the stage of the planning process when decisions amongst alternative scenarios are to be made. Specifically, it addressed the issue of brownfield regeneration in the city of Trento, where the presence of thirteen sites that could be converted to new public green areas determines the need for a rational approach to select and prioritise interventions. The presence of brownfields is a key issue for today's cities, with strong economic and social implications (Nassauer and Raskin 2014). Recent studies have highlighted how brownfields can be turned into sources of ES for the urban population (McPhearson et al. 2013; Collier 2014; Mathey et al. 2015; Geneletti et al. 2016; Beames et al. 2018), thus contributing to a more sustainable urban development (European Commission 2016).

Our analysis focused on the expected benefits of brownfield regeneration scenarios in terms of improved cooling effect by vegetation and enhanced opportunities for nature-based recreation: two of the most critical issues for citizens' wellbeing in Trento. The results of the two ES assessments were expressed through beneficiary-based indicators. Several authors have highlighted limitations and potential drawbacks associated to the use of both biophysical measures and monetary valuations of ES in real-life decision-making contexts (Bagstad et al. 2014; Ruckelshaus et al. 2015; Saarikoski et al. 2016; Olander et al. 2018). To this respect, beneficiary-based indicators that explicitly link the provision of ES with changes in human wellbeing are considered a promising way to integrate ES knowledge into decision-making processes (Geneletti et al. 2016; Olander et al. 2018) and to communicate ecological knowledge to planners and politicians primarily focused on social and economic objectives (von Haaren and Albert 2011; Schleyer et al. 2015), as is often the case in urban planning.

Multi-criteria analysis proved to be an effective tool to make the results of multiple ES assessments usable by decision-makers. On the one hand, it allows multiple sources of information and value dimensions to be combined, disregarding the indicators that are used to express them (Saarikoski et al. 2016). On the other hand, it offers a structured way to explore different stakeholder perspectives and related objectives, balancing diverse and sometimes competing interests in a rational and transparent way (Adem Esmail and Geneletti 2018). In the analysed case, all scenarios had a positive effect on the provision of both ES, but potential trade-offs could be related to competing uses of the brownfields (Kain et al. 2016) or to the costs of intervention (Koschke et al. 2012). Starting from the described multi-criteria analysis, a more complete decision support system could be built by integrating other relevant criteria about costs and benefits of planning scenarios.

As already demonstrated in other applications (Sanon et al. 2012; Grêt-Regamey et al. 2013; Kremer and Hamstead 2016), the results of the analysis show how priorities may shift with changing policy goals. The ranking is sensitive to the relative weights assigned to the two ES and related categories of beneficiaries and, eventually, the best scenario depends on decision-makers' orientations about specific planning objectives. This highlights the need for a strategic approach to ES, still mostly lacking in current planning documents (Cortinovis and Geneletti 2018b). Clear objectives and targets for ES provision would help to identify the values against which the effectiveness of planning actions should be measured, and to highlight the relevance of ES knowledge within the planning process. Information about ES beneficiaries allows understanding the social implications of planning decisions, particularly in terms of equity and environmental justice, as discussed in the next chapter.

Chapter 6
Towards Equity in the Distribution of Ecosystem Services in Cities

6.1 Introduction

A major challenge for cities worldwide is achieving an equitable distribution of urban services (UN-Habitat 2016). Among the latter, urban ecosystem services (ES) are increasingly acknowledged for their role in contributing to health and wellbeing, hence promoted as a valid nature-based solution to many urban challenges (European Commission 2015). Yet, urban ecosystems, such as parks, trees, gardens, greenways, are heterogeneously distributed over space, and so are the ES they provide, which may cause inequality in the distribution of benefits to citizens (Ernstson 2013). Through urban planning, local administrations can manage the distribution of urban ecosystems and their services in a city, determining the number, location, and type of beneficiaries they reach (Kremer et al. 2013). However, this requires moving beyond general urban quality standards, such as per capita green space targets (Badiu et al. 2016; Kabisch et al. 2016), which do not capture details about the actual distribution of benefits across different areas and population groups (Larondelle and Haase 2013; Cortinovis et al. 2018).

McDermott and colleagues identified three dimensions of equity in relation to the local provision of ES, namely distributional equity, procedural equity, and contextual equity (McDermott et al. 2013). Distributional equity considers the allocation of benefits and costs associated to ES provision among stakeholders. Procedural equity looks at the level of participation and representation in decision-making processes that affect the provision of ES and related benefits. Contextual equity refers to the conditions that determine people's capability to participate to decision-making processes, as well as to access ES benefits, including physical, socioeconomic, and institutional barriers.

The vast literature on ES mapping is well equipped to assess distributional equity, possibly accounting for relevant contextual factors. A number of methods in the biophysical, socio-cultural, and economic domains exist to understand how supply

© The Author(s) 2020
D. Geneletti et al., *Planning for Ecosystem Services in Cities*, SpringerBriefs in Environmental Science, https://doi.org/10.1007/978-3-030-20024-4_6

and demand of ES, and related benefits and values, vary across space (Harrison et al. 2017; Santos-Martin et al. 2018). However, there are key elements that ES assessments need to consider, to be informative about the equitable distribution of ES. This chapter explores these elements, presents a case study application, and provides some conclusions on how planning can pursue equitable distribution of ES in cities.

6.2   Elements to Assess Equity in the Distribution of ES

Three spatial elements are important to understand the distributional equity of ES: supply (i.e., where ES services are produced), access (i.e., where and by whom ES are used), and demand (i.e., who needs ES). The first obvious step of the analysis consists in identifying where ES are actually supplied (Burkhard et al. 2012). To this end, many assessments rely on the spatial distribution of urban parks or public green areas as a proxy of the spatial distribution of ES supply in the city. In this way, however, they fail to account for the ES provided by other types of urban green areas, including private gardens (Lin et al. 2017), street trees and roadside vegetation (Mullaney et al. 2015; Säumel et al. 2016), community gardens (Camps-Calvet et al. 2015), or even brownfields and abandoned areas (Mathey et al. 2015; Pueffel et al. 2018).

Moreover, spatial analyses focused on the distribution of green areas in the city sometimes overlook the fact that different ecosystems supply different types of ES and with different levels of effectiveness (de Groot et al. 2010). As Daw et al. 2011 put it, assuming all green areas as "black boxes" providing ES – without being able to determine which ES are supplied, and to which level of effectiveness – is a poor starting point to address issues of equity. Worth of mention here is the fact that ES indicators should relate to an appropriate scale and resolution. While more and better data are generally available in cities compared to non-urban areas, informing urban planning decisions requires capturing changes with a high level of detail; hence, data availability may sometime represent a crucial barrier. To effectively capture ES supply and provide useful information for equitable distribution of ES, the assessment should consider all ecosystems in a city, not only public green spaces. In addition, it should provide disaggregated information about the different ES that are supplied, considering the extent to which different urban ecosystems contribute to the provision of each ES.

Comparing ES supply and demand in a spatially explicit way is a fundamental step to understand who benefits from which services (Burkhard et al. 2012). However, simple methods based on spatial overlays may overlook whether the potential beneficiaries can actually access the supply of ES (Schröter et al. 2012). Tallis and Polasky (2009) rightly refer to access to ES, which is different from access to ecosystems or to public green spaces. Areas of ES generation (*service providing areas*) do not always correspond to the areas where ES can be actually enjoyed (*service benefitting areas*) (Syrbe and Walz 2012). This is especially evident when the two areas are not physically connected, as in the case of food and raw materials that are traded across the globe, but it applies also for many regulating ES, including pollination, micro-climate regulation, flood protection, and noise mitigation (Fisher et al. 2009). Characterizing the flow of ES in terms of spatial relation between providing and benefitting areas is therefore instrumental to understand where the effects of planning actions are to be expected, hence to identify the associated "winners" and "losers" (Schröter et al. 2012).

While benefitting areas provide a good basis to measure access to ES, contextual factors restricting the ability of people to enjoy the benefits derived from ES should also be taken into account. These may include physical barriers, as well as formal (i.e., laws and regulations) and informal (e.g., social norms and cultural practices) institutional barriers that limit the access to ES. For example, land tenure can be used as a proxy to assess formal institutional access, since it provides a basis to discriminate between different potential users of public, common, and private spaces. Other aspects, including overcrowding, perception of safety, social bonds, and community cohesion are more difficult to capture, but can be relevant in determining who actually benefit from ES (Lee et al. 2015). In conclusion, to determine whether access to ES occurs or not, first, it is crucial to define the flow of ES and the spatial relation between providing and benefitting areas. Then, it is important to capture the possible physical and institutional barriers that may limit the access to benefitting areas, possibly identifying the affected population groups.

Concerning demand, a common practice in ES assessment is to consider it as equally distributed among all citizens, either by fixing standard thresholds (Kabisch and Haase 2014) or by relying on population density as a proxy (Baró et al. 2016), disregarding demographic and socio-economics differences among individuals and population groups. Yet, equity does not require everybody accessing the same types and amount of ES. On the opposite, it should be based on the analysis of the specific needs of each individual or population group (McDermott et al. 2013). Different people derive benefits from different ES, hence may express differentiated needs (Rodríguez et al. 2006). Understanding "who counts for equity" thus requires measuring the need for ES of different individuals and population groups by comparing their capabilities, the costs, benefits, risks, and opportunities associated to ES provision, as well as factoring variables like gender, age, and health status (Sen 2009).

Besides the "who", it is equally important to define "where" such demand is located (McDermott et al. 2013). Methods for mapping ES demand, i.e. describing how it varies across space, are manifold (Wolff et al. 2015). Applying them to a spatial analysis of the key variables that affect the demand of different individuals and population groups provides a disaggregated view that is essential to inform decision-making. Table 6.1 provides a summary of the elements described in this section.

Table 6.1 Some recommendations for ES assessment addressing equity in the distribution of ES

Supply	Access	Demand
(i) Consider all ecosystems in the city, not only public green spaces.	(i) Consider access to ES, rather than to ecosystems or public green spaces.	(i) Consider the "who" – the target for demand.
(ii) Provide disaggregated information about the supply of different ES.	(ii) Characterize the flow of ES in terms of spatial relation between providing and benefitting areas, direction, and scale.	(ii) Consider where the demand is located.
(iii) Specify to which extent different ecosystems contribute to the supply of different ES.	(iii) Identify physical and institutional barriers that may limit the access to ES benefitting areas.	

6.3   Application to a Study Area in the City of Trento

A test application of the elements presented in the previous section has been conducted for different ES in the city of Trento (see Chap. 5 for more information about the city). The assessment focused on four urban ES relevant in the local context. Four 500-by-500 m sample areas were selected, roughly corresponding to four different residential neighbourhoods (see Fig. 6.1). The four sample areas are located in the central part of the city, where population density is higher and presence of public green spaces more limited than in the surrounding urban space, and are characterized by different socio-demographic conditions. The total extension of urban ecosystems, including trees, grass and shrubs, and bare soil in the four sample areas covers almost 300.000 m^2, i.e. between one fourth and one third of the total surface. The average *per capita* amount of green space is around 50 m^2.

To compare the sample areas, four regulating ES were selected, namely, carbon storage, air pollution removal, micro-climate regulation (cooling), and noise mitigation. The four ES differ significantly in terms of ecosystem functions that support the provision, spatial relation between providing and benefitting areas, and type of demand, hence they provide an interesting testbed to apply the elements described in the previous section.

6.3.1   Assessing ES Supply

The assessment of the supply of the selected ES was based on average values for different types of urban ecosystems retrieved from previous studies in urban contexts (McPhearson et al. 2013; Derkzen et al. 2015; Zardo et al. 2017). Table 6.2 summarizes the selected values harmonized with respect to four main land cover types; namely, (i) built-up and sealed, (ii) bare soil, (iii) grass and shrubs, (iv) trees and woodland. More in detail, carbon storage was estimated for coarse vegetation (i.e. trees and woodland), fine vegetation (i.e., grass and shrubs), and soil adopting

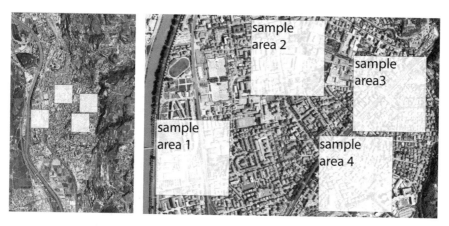

Fig. 6.1 The four selected sample areas in the city of Trento

Table 6.2 Average values for ES supply for different land cover types, in dimensional and dimensionless form

	Carbon storage		Air pollution removal		Micro-climate regulation		Noise reduction	
Built-up and sealed	–	0	–	0	–	0	–	0
Bare soil	8.2 kg/m²	5.3	–	0	1.2 °C	3.3	–	0
Grass and shrubs	8.4 Kg/m²	5.4	1.12 g/m²/year	4	1.2 °C	3.3	0.375 Db(A)/100m²	1.8
Trees and woodland	15.5 kg/m²	10	2.73 g/m²/year	10	3.6 °C	10	2 Db(A)/100m²	10

the same average values applied by McPhearson et al. (2013) to the city of New York. The same approach was used to assess air pollution removal, focusing on PM10 deposition on grass and woody vegetation (McPhearson et al. 2013). Based on Derkzen et al. (2015), values for air pollution removal where doubled for green areas located within a 50-m buffer from streets to account for the higher concentration of PM10 that increases the deposition flux. Cooling was assessed following the approach described in Chap. 4. Finally, noise reduction was assessed by adopting the values proposed by Derkzen et al. (2015). For each ES, the values were then converted into dimensionless scores ranging from 0 to 10, where 10 corresponds to the value of the best-performing land cover type.

Operationally, the first step to map the supply of ES consisted of the visual inspection of an aerial image to identify the main land cover types in the four sample areas, followed by screen digitising in a GIS. The land-cover information was then combined with the standardized values from Table 6.2 to obtain supply maps for each ES. To get an overall indicator of ES supply in each sample area, the surface of each land-cover type was multiplied by the corresponding standardized value, and then the contributions of the different land cover types were summed. Finally, the result was divided by the total surface of the sample area to obtain an overall score ranging from 0 to 10, which measures the supply of each ES in each sample area.

Figure 6.2 shows the results of the supply analysis. Overall, Sample area 3 shows the highest supply score for three of the four analysed ES, namely: air pollution removal, cooling, and noise reduction. For carbon storage, the best performance is represented by Sample area 1. Noteworthy is that Sample area 3 performs better despite the lower amount of green areas compared to Sample area 1. Furthermore, the performance of each sample area varies depending on the ES that is being considered. This shows the importance of the typology of green areas in determining the supply of ES, and the potential bias that aggregated indicators (such as per-capita green area) can introduce in the representation of the equitable distribution of ES.

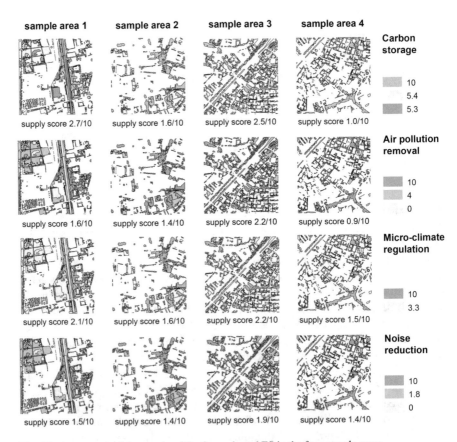

Fig. 6.2 Assessment of the supply of the four selected ES in the four sample areas

6.3.2 Assessing Access to ES

The assessment of the access to the four selected ES considered two main aspects, mentioned in Table 6.1. First, service benefitting areas were identified based on the flow of ES, by considering the specific spatial relation between providing and benefitting areas, direction, and scale that characterize each ES. Second, the presence of physical or institutional barriers to access were verified in each case. The relevant spatial characteristics of ES flow, the potential barriers to access, and the method adopted is presented in Table 6.3.

In the case study, given the small size of the samples, it was assumed that all benefitting areas are physically accessible, i.e., within walking distance from households (Kabisch and Haase 2014). For the ES to which institutional access might be relevant, land tenure was used as a proxy to restrict the analysis to benefitting areas accessible to all residents. Hence, only public areas are considered as possible places for enjoying the benefits of cooling and noise reduction, while carbon storage and air pollution removal are not affected by any limitation. The overall access score for each sample area was obtained by applying the same procedure used for the

Table 6.3 Relevant spatial characteristics of ES flow, potential barriers to access, and method adopted to assess access for the four selected ES

	ES flow and characteristics of benefitting areas	Method
Carbon storage	Global and non-proximal, equally distributed across the city; access to specific areas is not required to benefit from the ES	All sample areas are assigned the same access score calculated as the average of the supply scores of the four sample areas. The score is considered as homogeneously distributed over the whole study area.
Air pollution removal	Local (district) and non-proximal, evenly distributed across the neighbourhood; access to specific areas is not required to benefit from the ES	For each sample area, the score calculated in the supply analysis is used. The score is considered as homogeneously distributed over each sample area.
Micro climate regulation	Local (site/block) and proximal, the distribution is considered omnidirectional with benefitting areas up to some hundreds meters all around the area. Access to defined benefitting areas is required to benefit from the ES.	For green areas with an extension of less than 2 ha, the estimated cooling effect is perceivable up to a distance of about 100 m from the site Shashua-Bar and Hoffman (2000). Those areas are assigned a score that is half of that of the green area itself. Moreover, to consider the local contribution of shading by trees, a buffer of 5 m around the canopies is added with the same score as the supply score assigned to trees.
Noise mitigation	Local (site) and proximal, the distribution is directional with benefitting areas up to some tens meters from the noise source. Access to defined benefitting areas is required to benefit from the ES.	Benefitting areas are identified by creating a 50 m-buffer from the streets and their score is set as half of the supply score of the green area that provides the service Derkzen et al. (2015).

supply, resulting in a normalized score ranging from 0 to 10. The final access score also incorporates the assessment of ES supply conducted in the previous step, hence the access score could also be interpreted as a 'reduced' supply score that accounts for issues related to the spatial distribution of service benefitting areas and their accessibility, when relevant to the specific ES under investigation.

The results of the access analysis vary depending on the ES. For carbon storage and air pollution removal, which are assumed to be equally distributed over the sample areas, no further spatial analysis is needed. For carbon storage, the same score is assigned to all areas, while for air pollution removal a score equal to the supply score is used to characterise the different sample areas. For the ES that produce local benefitting areas characterised by proximity to the providing areas, the results of the access analysis are shown in Fig. 6.3. For both micro-climate regulation and noise reduction, the ranking of the sample areas based on the access score is different than the ranking based on the supply score

6.3.3   Assessing ES Demand

According to Wolff et al. (2015), depending on the type of ES, the demand can be assessed either based on direct use, or based on preferences for a desirable level of ES supply. In the case of regulating services, many assessments adopts the latter method, defining the desirable level through indicators of vulnerability (Wolff et al. 2015). Here we refer to Kazmierczak (2012), who identified four main vulnerable groups based on criteria of poverty, diversity (presence of foreigners), and age, specifically distinguishing children (0–4 years old) and elderly (above 65 years old). In the case study, the poverty indicator was not considered due to the lack of disaggregated spatial data. The other three criteria were assessed based on census data provided by the local administration. The number of residents, children, elderly, and

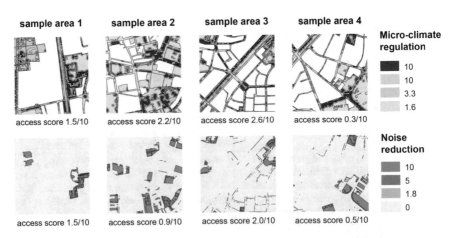

Fig. 6.3 Assessment of access to two proximity-dependent ES in the four sample areas

foreigners for each census block was linked to the map by considering all groups as evenly distributed on the surface covered by the footprint of residential buildings. Noteworthy, overlapping of vulnerabilities (e.g. an old person that is also a foreigner) were not taken into account; instead, any vulnerable was counted as one unit assuming that an individual that is both old and foreigner and the presence of one old person and one foreigner would raise equally the vulnerability of the neighbourhood. The overall score of ES demand was obtained by normalizing the number of vulnerable individuals in the sample areas on a scale between 0 and 10. The distribution of vulnerable individuals is almost proportional to the distribution of the population density, i.e. the area with the highest population density is also the area with highest number of vulnerable individuals.

6.3.4 Combining Information of Supply, Access and Demand

Taken singularly, the results of the analyses described in the previous sub-sections can be used to identify hotspots in the city where supply, access and demand for ES are particularly high. The supply, access and demand scores can also be aggregated to unveil, for example, mismatches between demand and supply (Ortiz and Geneletti 2018). As an example of aggregation, we proposed a combined indicator based on the scores of access and demand analysis, given the former also include the supply score. Operationally, for each ES and each sample area, the access score was divided by that of demand, so that higher values correspond to better performances. The combined indicator allows ranking different areas of the city, based on a comparison between the existing demand for ES, and their actual availability for the citizens. Hence, it can provide more information with respect to considering only supply and demand. As an illustration, Fig. 6.4 compares this indicator with assessments based on supply only, and the ratio between supply and demand (without considering access). The comparison focuses on cooling and noise reduction, two examples of ES for which access is a relevant factor to consider in the assessment.

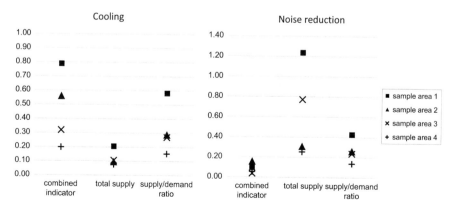

Fig. 6.4 Comparison of the combined indicators with the total supply and the ratio between supply and demand in the four sample areas for cooling (left) and noise reduction (right)

6.4 Conclusions

An equitable distribution of resources, and more specifically of ES, is one of the key elements in the purse of equity, and sustainability. ES assessments that investigate the distribution of ES can play an important role in addressing equity issues in urban planning. The concept of ES represents a very useful tool since it allows understanding and quantifying how ecosystems provide services and spatially defining the relationship between their structure, functions, and the related benefits at a suitable scale (Braat and de Groot 2012). In this chapter, we suggest that ES assessments need to capture information about the supply, access, and demand of services in a city, in order to provide planners with useful information on the degree of equity in the distribution of ES.

The approach described in the case study application present limitations. Firstly, access to ES was estimated in a simplified way, reflecting the limited contributions on this topic found in the literature. There is a need for studies that go more in depth in unveiling, and mapping all the factors that may play a role in determining how ES are accessed by beneficiaries. Secondly, demand was estimated by considering only few categories of vulnerable individuals. Other socio-economic and cultural factors (e.g., housing conditions) could be added to improve the analysis, exploring the complexity of the demand for regulating ES in cities, and distinguishing among different groups of beneficiaries associated to different ES. Thirdly, the assessment of supply was also based on one single indicator for each ES. Using multiple indicators could provide a better picture of different types of supply, reflecting for example different potential uses by beneficiaries.

In conclusion, the potential of applying holistic approaches to the investigation of distribution of ES, which include biophysical, as well as socio-economic considerations, cannot be ignored. Despite using disaggregated data and complex assessments can be costly and time-consuming (Gómez-Baggethun and Barton 2013), there is a need to keep on walking this path to improve the outcome of planning processes.

Chapter 7
Conclusions

7.1 Summary of the Main Messages

This book focused on the relation between urban planning and ecosystem services (ES), acknowledging their potential role in addressing many challenges of todays' cities. Planning decisions are one of the most influential factors determining the amount and spatial distribution of both ES supply and demand in cities. Hence, there is a need to integrate ES knowledge in planning processes, not only to measure and possibly reduce the negative impacts of planning decision on ES provision, but also to enable the proactive enhancement of ES through effective actions. To this overall aim, the book reviewed the state of the art of ES integration in current planning practices (Chaps. 2 and 3), presented an exemplary model for ES assessment specifically developed to support urban planning (Chap. 4), illustrated how ES information can be applied to support real-life planning decisions (Chap. 5), and discussed the design of ES assessments to analyse the equitable distribution of ES in cities (Chap. 6).

Chapter 2 presented the results of a review of 22 comprehensive spatial plans of Italian cities, while Chap. 3 focused on a more recent type of planning documents, i.e. climate adaptation plans, reviewing a set of 14 plans of European cities. By looking at how ES are actually addressed in the different plan components, disregarding the terminology used to refer to them, the reviews revealed what ES knowledge is already included in current plans and what gaps are still to be filled. Interestingly, the two reviews present some common results that suggest an overall trend in the way ES knowledge is being mainstreamed and taken-up among practitioners and decision-makers. Both reviews highlight a strong awareness of some ES and of the benefits they provide in terms of climate change adaptation (e.g., heatwave mitigation and flood control), and human health and wellbeing (e.g., recreation, air pollution reduction, noise mitigation). A high number of ES-related actions

© The Author(s) 2020
D. Geneletti et al., *Planning for Ecosystem Services in Cities*, SpringerBriefs in Environmental Science, https://doi.org/10.1007/978-3-030-20024-4_7

was found in both types of plan, and the analysis of spatial plans revealed a wide variety of tools for implementation.

However, the proposal of ES-related measures is rarely supported by an adequate knowledge base and analysis, which may eventually undermine their effectiveness. The general idea that "more green will do some good" seems to guide the inclusion of ES-based actions in current plans, where critical decisions about the design and the location of interventions are seldom justified by the analysis of the expected outcomes and the distribution and vulnerability of the potential beneficiaries. As such, the most common measure found in climate adaptation plan was the enhancement of green areas, a typical strategy that planners put in place for a variety of purposes that go beyond climate change adaptation. Moreover, while a set of ES, including regulating ES related to climate change adaptation, are widely acknowledged also in comprehensive spatial plans, others are hardly considered. This may lead to trade-offs unconsciously generated by planning decisions and, ultimately, to a loss of important but underestimated ES.

Chapter 4 illustrated how an ES model can be developed with the aim of supporting urban planning decisions. Specifically, it describes the development of a spatially-explicit model to map and assess micro-climate regulation provided by different typologies of urban green infrastructures. The model is based on an extensive review of the scientific literature, and summarises the main findings in a way that is accessible and usable by planners. Tree canopy coverage, soil cover, and size are identified as the most relevant variables determining the cooling potential of urban green infrastructure. By combining the three variables, 50 typologies of urban green infrastructure are defined, and each of them is assigned a score depending on the climatic region of interest. Planners can directly refer to these "archetype" typologies to assess the existing condition of a city based on commonly available data (as demonstrated by the application to Amsterdam), as well as to measure the expected benefits of planning interventions (as exemplified in the case study in the city of Trento presented in Chap. 5).

Chapter 5 moved a step further in the operationalisation of ES knowledge, showing how the information produced by ES mapping and assessment can be used to support real-life planning decisions. The case study in Trento (Italy) demonstrated that the assessment of ES and related benefits can be adopted to prioritise planning scenarios, as for the example of brownfield regeneration. The case study considered two illustrative but critical ES for the context, i.e. micro-climate regulation and nature-based recreation, applying models specifically developed (as the one described in Chap. 4) or adapted (as is the case of ESTIMAP-recreation) for urban planning purposes. Combining the assessment of ES supply with spatial information on the potential beneficiaries and their different levels of vulnerability proved to be an effective way to build a common ground between ES assessments and urban planning. A multi-criteria analysis offered a structured way to combine multiple indicators in a synthetic and usable outcome for decision-makers, while exploring and balancing different stakeholder perspectives and potentially competing interests in a rational and transparent way.

Finally, Chap. 6 focused on an emerging issue in the urban ES science with relevant implications for urban planning, i.e. the analysis of equity in the distribution of ES and related benefits across the city. While this is still an innovative field where the ES concept is being applied, some key requirements for ES assessments to address the equitable distribution of ES in cities can be already identified. Specifically, access emerges as a fundamental aspect that shapes the flow of ES by determining the matching between supply and demand. To capture the real distribution of ES across areas and population groups, ES assessments should take into account the barriers, including both physical and institutional barriers, that may limit the access to ES. A spatially-explicit ES assessment that considers how ES flow from providing areas (characterised in terms of their specific ES potential), to benefitting areas (characterised by a certain level of demand and accessibility) is therefore essential to support urban planning in pursuing equity in the distribution of ES in cities.

7.2 Challenges for Future Research and Practice

Despite the explicit link with decision-making being among its distinctive traits since its origin (Daily et al. 2009), the integration between ES science and planning practices is still limited, especially at the urban scale. In this context, the 'salience' or 'relevance' of ES knowledge, a key attribute to measure its capacity to inform decision-making (Cash et al. 2003), corresponds to the extent to which it responds to knowledge needs that are not yet answered by the set of concepts, approaches, and methods already adopted in current urban planning practices. Understanding the specific needs and requirements that urban planning poses to ES knowledge is therefore fundamental for ES research. To this aim, the reviews of planning documents presented in Chaps. 2 and 3 revealed some shortcomings in how are ES addressed and suggest some entry points where ES knowledge could be profitably integrated.

First, a need for enhancing the knowledge base on ES clearly emerged. This can be partly ascribed to gaps in ES research: an appropriate knowledge base is fundamental to design effective actions, but requires adequate methods that, in some cases, are still not available. The set of ES less frequently mentioned in urban plans are also the least popular in the ES literature, which may indicate that the respective functions are well known from the ecological point of view, but still not acknowledged for their contribution to human wellbeing. Moreover, methods to assess many urban ES require further efforts to be tailored to urban planning needs, in two respects. On the one hand, urban contexts, characterized by heterogeneity and fragmentation of green infrastructure components, define specific requirements for biophysical methods in terms of accuracy and resolution, which limits the transferability of methods developed at different spatial scales. On the other hand, the integration in planning processes determines data and resource constraints (e.g., in terms of time, costs, and expertise). Methods able to capture the specificity of urban ecosystems while meeting the requirements of planning applications are needed, such as the model presented in Chap. 4. The steps that were followed to analyse and synthesize the scientific literature from multiple disciplines, and the type of results proposed could serve as an inspiration to develop other ES methods usable by planners.

However, biophysical indicators are not sufficient to communicate ES values in a meaningful way, especially in decision-making processes mostly driven by social and economic objectives, as is often the case in urban planning. On the contrary, beneficiary-based indicators that explicitly link the provision of ES with changes in human wellbeing are a promising way to communicate ecological knowledge to planners and decision-makers. As demonstrated by the application described in Chap. 5, indicators describing ES values by accounting for their beneficiaries and the different levels of demand that they express have the capacity to reflect different planning objectives and stakeholders' perspectives, hence to inform and support planning decisions. In order to better assess the demand for ES, future applications can take advantage of approaches commonly adopted by urban planning (e.g., spatial analysis of population, identification of specific target groups, elicitation of citizens' preferences and opinions), and further develop their use.

Combining values associated to different ES is also a challenge. Scientific advancements produced by transdisciplinary efforts are needed to define appropriate and effective methods that allow combining the results of biophysical, sociocultural, and economic methods, thus accounting for the multiple values of ES. As shown in Chap. 5, a methodological support is provided by multi-criteria analysis techniques, which offer a platform for combining multiple value dimensions, integrating stakeholders' opinions along with technical inputs. Multi-criteria analysis allows exploring different perspectives and balancing competing interests to find an agreed-upon solution. Urban planning would benefit from adopting such methodologies, which enhance participation and transparency, ultimately strengthening the ownership of the results. However, a pre-requisite is the identification of clear ES-related objectives. Increasing ES supply is not equal to increasing the number of ES beneficiaries, which again is not the equal to maximising ES benefits produced by planning decisions. Consequently, only a clear definition of planning objectives can set the basis for the design of effective planning actions.

A better integration of ES in the strategic component of urban plans and the definition of objectives and targets for ES enhancement require further efforts by planners and decision-makers. While the high number of actions based on ES in the reviewed plans indicates that planners have the capacity to enhance the future provision of urban ES, strengthening the consideration of ES as a strategic issue is fundamental to ensure long-term commitment in the implementation and monitoring of planning action. A strategic approach to ES also guarantees that all relevant ES for the context are taken into account. Urban green infrastructure components act as providing units of multiple ES, hence planning actions can be expected to produce effects on a bundle of ES. This is not only the case of planning actions specifically aimed at enhancing ES provision or directly affecting urban green infrastructure, but it is equally true for planning actions affecting the distribution of beneficiaries, the environmental conditions of the city, or the accessibility to certain areas. While the multifunctionality of urban green infrastructure generates potential synergies, which are indicated as one of the main strengths of ecosystem-based actions, looking at the whole range of ES affected by planning actions may reveal trade-offs, hence unexpected and undesired outcomes of planning decisions.

A final challenge is related to the development and application of appropriate methods for assessing equity in the distribution of ES. Inequality and exclusion are rising in most cities in the world and indicators focused on the average wellbeing, overlooking its distribution, may hide an increasing divide between different population groups, thus leading to wrong conclusions about the impacts of policies and decisions. Methods for mapping ES, which describe how ES supply and demand, as well as related benefits and values, vary across space, are a good starting point to assess distributional equity, possibly accounting for relevant contextual factors. However, as discussed in Chap. 6, some criteria must be fulfilled to ensure that a spatial assessment of ES provides useful information to pursue an equitably distributed wellbeing and further advancements are still needed on this issue.

7.3   Concluding Remarks

This book moved from the belief that the ES approach, grounded on the growing ES science, can be a valuable support to improve decision-making towards planning more sustainable, liveable, and resilient cities. From this perspective, ES science is seen as a provider of credible and relevant knowledge that can complement the set of information – a combination of both scientific and traditional/local knowledge – on which planning decisions are usually based, thus promoting a stronger incorporation of ecological concerns into decision-making processes that mainly pursue socio-economic goals. Operationally, ES science offers to urban planners a wide range of methods and tools that can be used to analyse the current condition and to assess the expected impacts of planning decisions. In this context, particularly relevant are spatially-explicit methods that allow mapping the distribution of ES and related benefits across the city.

As suggested by our exploration, the expected benefits of a further integration of ES in urban planning are manifold. First, it supports the design of appropriate and effective actions, being they targeted to promote climate change adaptation or to enhance citizens' health and wellbeing. Second, it increases the awareness of a larger set of values that are at stake in planning processes, including values normally underrepresented, thus highlighting co-benefits and trade-offs that may arise from planning actions, and allowing for a more transparent and conscious prioritization. Third, it promotes the explicit identification of ES demand and beneficiaries, thus enhancing the consideration of equity issues and providing planners and decision-makers with stronger arguments against conflicting interests on land-use decisions.

What emerges is an overall coherence between the ES approach and urban planning objectives that aim at enhancing human wellbeing in cities. Hence, the integration does not require to reshape current practices, but it can build on what is already there, as also demonstrated by our reviews. Some promising fields of cross-fertilization between ES and urban planning emerged in this book, including possible contributions of the urban planning discipline to the ES science (Cortinovis 2018). For example, methods and approaches already in use by planners can be applied to identify actual and potential beneficiaries, to quantify ES benefits, and to analyse equity in ES provision. Implementation tools to support the operationalization of ES knowledge represent another key contribution. So far, ES science has developed very few models for implementation. On the contrary, urban planners are equipped with a large toolbox where suitable tools to implement ES-informed decisions and to secure and enhance the provision of ES in cities can certainly be found (e.g., regulations, building standards, financial and non-financial incentives, among others).

A growing demand for ES knowledge to be integrated in urban planning practices is determined by the strong support that ecosystem-based actions and nature-based solutions are receiving (European Commission 2015). However, as it emerged from the book, the ES approach, providing a holistic framework that describes the multiple relations between ecosystems and human wellbeing, offers to urban planning much more than solutions. Within this framework, objectives that account for the complex interactions between the ecological and the socio-economic spheres can be set, and decisions assessed based on their expected long-term consequences. Urban planning plays a key role in coordinating different sectoral policies and bridging multiple institutional scales, hence it can be the starting point for ES knowledge to permeate other decision-making processes. Urbanization is one of the major threats to biodiversity and ES worldwide. In this respect, promoting the ES approach through urban planning may seem a paradox, but it is also a great opportunity to make human development truly sustainable.

Appendix

Annex I – List of the Planning Documents Reviewed in Chapter 2

ID	City	Year of approval	Population 1/1/2014*	City area [km²]
P01	Ascoli Piceno	2014	50,079	158.02
P02	Benevento	2012	60,770	130.84
P03	Brescia	2012	193,599	90.34
P04	Como	2013	84,834	37.12
P05	Cremona	2013	71,184	70.49
P06	Genoa	2014	596,958	240.29
P07	Lecco	2014	48,131	45.14
P08	Mantua	2012	48,588	63.81
P09	Milan	2012	1,324,169	181.67
P10	Padua	2014	209,678	93.03
P11	Pavia	2013	71,297	63.24
P12	Piacenza	2014	102,404	118.24
P13	Prato	2013	191,268	97.35
P14	Rovigo	2012	52,099	108.81
P15	Savona	2012	61,761	65.32
P16	Treviso	2013	83,145	55.58
P17	Trieste	2014	204,849	85.11
P18	Udine	2012	99,528	57.17
P19	Varese	2014	80,927	54.84
P20	Venice	2014	264,534	415.90
P21	Vercelli	2012	46,992	79.78
P22	Vibo Valentia	2014	33,675	46.57

© The Author(s) 2020
D. Geneletti et al., *Planning for Ecosystem Services in Cities*, SpringerBriefs in Environmental Science, https://doi.org/10.1007/978-3-030-20024-4

Annex II – List of the Planning Documents Reviewed in Chapter 3

City	Name of the plan	Year	Source
Amsterdam	Amsterdam: a different energy (SEAP)	2010	http://mycovenant.eumayors.eu/
	Amsterdam definitely sustainable	2011	http://www.nieuwamsterdamsklimaat.nl/
	New Amsterdam climate	2010	http://mycovenant.eumayors.eu/
	Outspokenly sustainable-perspective 2014	2009	http://www.nieuwamsterdamsklimaat.nl/
	Structure vision for Amsterdam 2014	2008	http://www.nieuwamsterdamsklimaat.nl/
Barcelona	The energy, climate change and air quality plan for Barcelona (SEAP)	2011	http://mycovenant.eumayors.eu/
Berlin	Berlin Environmental Relief Programme (10 years) (SEAP)	2011	http://mycovenant.eumayors.eu/http://www.berlin.de/
Copenhagen	Copenhagen climate adaptation plan (SEAP)	2011	http://mycovenant.eumayors.eu/http://www.kk.dk/
Heidelberg	Climate protection commitment Heildelberg (SEAP)	2010	http://mycovenant.eumayors.eu/
London	Delivering London's energy future (SEAP)	2010	http://mycovenant.eumayors.eu/
	The London Plan: spatial development strategy for a greater London	2008	http://www.london.gov.uk
Madrid	Plan de uso sostenible de la energia y prevencion de cambio climatico (SEAP)	2008	http://mycovenant.eumayors.eu/
Milano	Piano per l'energia sostenibile ed il clima (SEAP)	2009	http://mycovenant.eumayors.eu/
Paris	Paris climate protection plan (SEAP)	2004	http://mycovenant.eumayors.eu/
Roma	Piano d'azione per l'energia sostenibile per la città di Roma (SEAP)	2010	http://mycovenant.eumayors.eu/
Rotterdam	Investing in sustainable growth, Rotterdam programme on (SEAP)	2010	http://mycovenant.eumayors.eu/
	Rotterdam climate city, mitigation action programme	2010	http://www.rotterdamclimateinitiative.nl/
	The new Rotterdam, Rotterdam climate initiative	2009	http://www.rotterdamclimateinitiative.nl/
Stockholm	Stockholm action plan for climate and energy (SEAP)	2012	http://mycovenant.eumayors.eu/ http://www.stockholm.se/
	Stockholm climate initiative	2010	http://www.stockholm.se/
Venezia	Piano d'azione per l'energia sostenibile (SEAP)	2013	http://mycovenant.eumayors.eu/
Warsaw	Sustainable action plan for energy Warsaw (SEAP)	2011	http://mycovenant.eumayors.eu/

References

Adem Esmail B, Geneletti D (2018) Multi-criteria decision analysis for nature conservation: a review of 20 years of applications. Methods Ecol Evol 2018:42–53. https://doi.org/10.1111/2041-210X.12899

Akbari H, Davis S, Dorsano S et al (1992) Cooling our Communities. A Guidebook on Tree Planting and Light-Colored Surfacing, Washington, DC

Allen RG, Pereira LS, Raes D, Smith M (1998) Crop evapotranspiration-Guidelines for computing crop water requirements. Food and Agriculture Organization of the United States, Rome

Andersson E, McPhearson T, Kremer P et al (2015) Scale and context dependence of ecosystem service providing units. Ecosyst Serv 12:157–164. https://doi.org/10.1016/j.ecoser.2014.08.001

Babí Almenar J, Rugani B, Geneletti D, Brewer T (2018) Integration of ecosystem services into a conceptual spatial planning framework based on a landscape ecology perspective. Landsc Ecol 33(12):2047–2059. https://doi.org/10.1007/s10980-018-0727-8

Badiu DL, Iojă CI, Pătroescu M et al (2016) Is urban green space per capita a valuable target to achieve cities' sustainability goals? Romania as a case study. Ecol Indic 70:53–66. https://doi.org/10.1016/j.ecolind.2016.05.044

Bagstad KJ, Villa F, Batker D et al (2014) From theoretical to actual ecosystem services: mapping beneficiaries and spatial flows in ecosystem service assessments. Ecol Soc 19. https://doi.org/10.5751/ES-06523-190264

Baker I, Peterson A, Brown G, McAlpine C (2012) Local government response to the impacts of climate change: an evaluation of local climate adaptation plans. Landsc Urban Plan 107:127–136. https://doi.org/10.1016/j.landurbplan.2012.05.009

Barbosa O, Tratalos JA, Armsworth PR et al (2007) Who benefits from access to green space? A case study from Sheffield, UK. Landsc Urban Plan 83:187–195. https://doi.org/10.1016/j.landurbplan.2007.04.004

Baró F, Palomo I, Zulian G et al (2016) Mapping ecosystem service capacity, flow and demand for landscape and urban planning: a case study in the Barcelona metropolitan region. Land use policy 57:405–417. https://doi.org/10.1016/j.landusepol.2016.06.006

Barton DN, Kelemen E, Dick J et al (2018) (Dis) integrated valuation – Assessing the information gaps in ecosystem service appraisals for governance support. Ecosyst Serv 29:529–541. https://doi.org/10.1016/j.ecoser.2017.10.021

Basu R, Samet JM (2002) Relation between elevated ambient temperature and mortality: a review of the epidemiologic evidence. Epidemiol Rev 24:190–202. https://doi.org/10.1093/epirev/mxf007

Bauler T, Pipart N (2013) Ecosystem services in Belgian environmental policy making: expectations and challenges linked to the conceptualization and valuation of ecosystem services. Elsevier

Beames A, Broekx S, Schneidewind U et al (2018) Amenity proximity analysis for sustainable brownfield redevelopment planning. Landsc Urban Plan 171:68–79. https://doi.org/10.1016/j.landurbplan.2017.12.003

Beery T, Stålhammar S, Jönsson KI et al (2016) Perceptions of the ecosystem services concept: opportunities and challenges in the Swedish municipal context. Ecosyst Serv 17:123–130. https://doi.org/10.1016/j.ecoser.2015.12.002

Berke PR, Godschalk D (2009) Searching for the good plan: a meta-analysis of plan quality studies. J Plan Lit 23:227–240. https://doi.org/10.1177/0885412208327014

Berndtsson JC (2010) Green roof performance towards management of runoff water quantity and quality: a review. Ecol Eng 36:351–360. https://doi.org/10.1016/j.ecoleng.2009.12.014

Bolund P, Hunhammar S (1999) Ecosystem services in urban areas. Ecol Econ 29:293–301. https://doi.org/10.1016/S0921-8009(99)00013-0

Botzat A, Fischer LK, Kowarik I (2016) Unexploited opportunities in understanding liveable and biodiverse cities. A review on urban biodiversity perception and valuation. Glob Environ Chang 39:220–233. https://doi.org/10.1016/j.gloenvcha.2016.04.008

Bouwma I, Schleyer C, Primmer E et al (2017) Adoption of the ecosystem services concept in EU policies. Ecosyst Serv. https://doi.org/10.1016/j.ecoser.2017.02.014

Bowler DE, Buyung-Ali LM, Knight TM, Pullin AS (2010a) A systematic review of evidence for the added benefits to health of exposure to natural environments. BMC Public Health 10:456. https://doi.org/10.1186/1471-2458-10-456

Bowler DE, Buyung-Ali LM, Knight TM, Pullin AS (2010b) Urban greening to cool towns and cities: a systematic review of the empirical evidence. Landsc Urban Plan 97:147–155. https://doi.org/10.1016/j.landurbplan.2010.05.006

Braat L, de Groot R (2012) The ecosystem services agenda: bridging the worlds of natural science and economics, conservation and development, and public and private policy. Ecosyst Serv 1:4–15. https://doi.org/10.1016/j.ecoser.2012.07.011

Brink E, Aalders T, Ádám D et al (2016) Cascades of green: a review of ecosystem-based adaptation in urban areas. Glob Environ Chang 36:111–123. https://doi.org/10.1016/j.gloenvcha.2015.11.003

Burkhard B, Kroll F, Nedkov S, Müller F (2012) Mapping ecosystem service supply, demand and budgets. Ecol Indic 21:17–29. https://doi.org/10.1016/j.ecolind.2011.06.019

Burkhard B, Maes J, Potschin-Young M et al (2018) Mapping and assessing ecosystem services in the EU – Lessons learned from the ESMERALDA approach of integration. One Ecosyst 3:e29153. https://doi.org/10.3897/oneeco.3.e29153

Burns MJ, Fletcher TD, Walsh CJ et al (2012) Hydrologic shortcomings of conventional urban stormwater management and opportunities for reform. Landsc Urban Plan 105:230–240. https://doi.org/10.1016/j.landurbplan.2011.12.012

Cadenasso ML, Pickett STA, Schwarz K (2007) Spatial heterogeneity in urban ecosystems: reconceptualizing land cover and a framework for classification. Front Ecol Environ 5:80–88

Cameron RWF, Blanuša T, Taylor JE et al (2012) The domestic garden – Its contribution to urban green infrastructure. Urban For Urban Green 11:129–137. https://doi.org/10.1016/j.ufug.2012.01.002

Camps-Calvet M, Langemeyer J, Calvet-Mir L, Gómez-Baggethun E (2015) Ecosystem services provided by urban gardens in Barcelona, Spain: insights for policy and planning. Environ Sci Policy 62:14–23. https://doi.org/10.1016/j.envsci.2016.01.007

Cao X, Onishi A, Chen J, Imura H (2010) Quantifying the cool island intensity of urban parks using ASTER and IKONOS data. Landsc Urban Plan 96:224–231. https://doi.org/10.1016/j.landurbplan.2010.03.008

Cash DW, Clark WC, Alcock F et al (2003) Knowledge systems for sustainable development. Proc Natl Acad Sci USA 100:8086–8091. https://doi.org/10.1073/pnas.1231332100

Castleton HF, Stovin V, Beck SBM, Davison JB (2010) Green roofs; building energy savings and the potential for retrofit. Energy Build 42:1582–1591. https://doi.org/10.1016/j.enbuild.2010.05.004

CBD (2011) Strategic plan for biodiversity 2011–2020. Including Aichi Biodiversity Targets, Montreal

Chang C-R, Li M-H, Chang S-D (2007) A preliminary study on the local cool-island intensity of Taipei city parks. Landsc Urban Plan 80:386–395. https://doi.org/10.1016/j.landurbplan.2006.09.005

Collier MJ (2014) Novel ecosystems and the emergence of cultural ecosystem services. Ecosyst Serv 9:166–169. https://doi.org/10.1016/j.ecoser.2014.06.002

Collier MJ, Nedović-Budić Z, Aerts J et al (2013) Transitioning to resilience and sustainability in urban communities. Cities 32. https://doi.org/10.1016/j.cities.2013.03.010

Conti S, Meli P, Minelli G et al (2005) Epidemiologic study of mortality during the Summer 2003 heat wave in Italy. Environ Res 98:390–399. https://doi.org/10.1016/j.envres.2004.10.009

Cortinovis C (2018) Integrating ecosystem services in urban planning. Doctoral Thesis. University of Trento

Cortinovis C, Geneletti D (2018a) Mapping and assessing ecosystem services to support urban planning: a case study on brownfield regeneration. One Ecosyst 3:e25477. https://doi.org/10.3897/oneeco.3.e25477

Cortinovis C, Geneletti D (2018b) Ecosystem services in urban plans: what is there, and what is still needed for better decisions. Land use policy 70:298–312. https://doi.org/10.1016/j.landusepol.2017.10.017

Cortinovis C, Zulian G, Geneletti D (2018) Assessing nature-based recreation to support urban green infrastructure planning in Trento (Italy). Land 7:112. https://doi.org/10.3390/land7040112

Costanza R, D'Age R, de Groot R et al (1997) The value of the world's ecosystem services and natural capital. Nature 387:253–260

Daily GC (1997) Nature's services. Societal dependence on natural ecosystems. Island Press, Washington, DC

Daily GC, Polasky S, Goldstein J et al (2009) Ecosystem services in decision making: time to deliver. Front Ecol Environ 7:21–28. https://doi.org/10.1890/080025

Daw TM, Brown K, Rosendo S, Pomeroy R (2011) Applying the ecosystem services concept to poverty alleviation: the need to disaggregate human well-being. Environ Conserv 38:370–379. https://doi.org/10.1017/S0376892911000506

de Groot R, Alkemade R, Braat L et al (2010) Challenges in integrating the concept of ecosystem services and values in landscape planning, management and decision making. Ecol Complex 7:260–272. https://doi.org/10.1016/j.ecocom.2009.10.006

Demuzere M, Orru K, Heidrich O et al (2014) Mitigating and adapting to climate change: multifunctional and multi-scale assessment of green urban infrastructure. J Environ Manage 146:107–115. https://doi.org/10.1016/j.jenvman.2014.07.025

Derkzen ML, van Teeffelen AJA, Verburg PH (2015) REVIEW: quantifying urban ecosystem services based on high-resolution data of urban green space: an assessment for Rotterdam, the Netherlands. J Appl Ecol 52:1020–1032. https://doi.org/10.1111/1365-2664.12469

Díaz S, Demissew S, Carabias J et al (2015) The IPBES conceptual framework – connecting nature and people. Curr Opin Environ Sustain 14:1–16. https://doi.org/10.1016/j.cosust.2014.11.002

Dick J, Turkelboom F, Woods H et al (2017) Stakeholders' perspectives on the operationalisation of the ecosystem service concept: results from 27 case studies. Ecosyst Serv. https://doi.org/10.1016/j.ecoser.2017.09.015

EEA (2012) Urban adaptation to climate change in Europe. Challenges and opportunities for cities together with supportive national and European policies, Copenhagen

Ehrlich PR, Ehrlich AH (1981) Extinction: the causes and consequences of the disappearance of species. Random House Inc, New York

Elmqvist T, Setälä H, Handel SN et al (2015) Benefits of restoring ecosystem services in urban areas. Curr Opin Environ Sustain 14:101–108. https://doi.org/10.1016/j.cosust.2015.05.001

Elmqvist T, Gómez-Baggethun E, Langemeyer J (2016) Ecosystem services provided by urban green infrastructure. In: Potschin M, Haines-Young R, Fish R, Turner RK (eds) Routledge handbook of ecosystem services. Routledge, Oxford, pp 452–464

Ernstson H (2013) The social production of ecosystem services: a framework for studying environmental justice and ecological complexity in urbanized landscapes. Landsc Urban Plan 109:7–17. https://doi.org/10.1016/j.landurbplan.2012.10.005

Escobedo FJ, Adams DC, Timilsina N (2015) Urban forest structure effects on property value. Ecosyst Serv 12:209–217. https://doi.org/10.1016/j.ecoser.2014.05.002

ETC/BD (2006) The indicative map of European biogeographical regions: methodology and development. Paris

European Commission (2006) Halting the loss of biodiversity by 2010 – and beyond – Sustaining ecosystem services for human well-being (COM/2006/0216 final). Brussels

European Commission (2010) Our life insurance, our natural capital: an EU biodiversity strategy to 2020 (COM/2011/0244 final). Brussels

European Commission (2015) Towards an EU research and innovation policy agenda for nature-based solutions & re-naturing cities. Final report of the Horizon 2020 expert group on nature-based solutions and re-naturing cities. Brussels

European Commission (2016) Urban Agenda for the EU "Pact of Amsterdam." Amsterdam

Farrugia S, Hudson MD, McCulloch L (2013) An evaluation of flood control and urban cooling ecosystem services delivered by urban green infrastructure. Int J Biodivers Sci Ecosyst Serv Manag 9:136–145. https://doi.org/10.1080/21513732.2013.782342

Fischer EM, Schär C (2010) Consistent geographical patterns of changes in high-impact European heatwaves. Nat Geosci 3:398–403. https://doi.org/10.1038/ngeo866

Fisher B, Turner RK, Morling P (2009) Defining and classifying ecosystem services for decision making. Ecol Econ 68:643–653. https://doi.org/10.1016/j.ecolecon.2008.09.014

Frantzeskaki N, Kabisch N, McPhearson T (2016) Advancing urban environmental governance: understanding theories, practices and processes shaping urban sustainability and resilience. Environ Sci Policy 62:1–6. https://doi.org/10.1016/j.envsci.2016.05.008

Geneletti D (2011) Reasons and options for integrating ecosystem services in strategic environmental assessment of spatial planning. Int J Biodivers Sci Ecosyst Serv Manag 7:143–149. https://doi.org/10.1080/21513732.2011.617711

Geneletti D (2013) Ecosystem services in environmental impact assessment and strategic environmental assessment. Environ Impact Assess Rev 40:1–2. https://doi.org/10.1016/j.eiar.2013.02.005

Geneletti D (2016) Handbook on biodiversity and ecosystem services in impact assessment. Edward Elgar Publishing, Cheltenham/Northampton

Geneletti D, Zardo L (2016) Ecosystem-based adaptation in cities: an analysis of European urban climate adaptation plans. Land use policy 50:38–47. https://doi.org/10.1016/j.landusepol.2015.09.003

Geneletti D, Zardo L, Cortinovis C (2016) Promoting nature-based solutions for climate adaptation in cities through impact assessment. In: Geneletti D (ed) Handbook on biodiversity and ecosystem services in impact assessment. Edward Elgar Publishing, Cheltenham/Northampton, pp 428–452

Geneletti D, La Rosa D, Spyra M, Cortinovis C (2017) A review of approaches and challenges for sustainable planning in urban peripheries. Landsc Urban Plan 165:231–243. https://doi.org/10.1016/j.landurbplan.2017.01.013

Geneletti D, Scolozzi R, Adem Esmail B (2018) Assessing ecosystem services tradeoffs across agricultural landscapes in a mountain region. Int J Biodivers Sci Ecosyst Serv Manag 14:1–35. https://doi.org/10.1080/21513732.2018.1526214

Gill S, Handley J, Ennos A, Pauleit S (2007) Adapting cities for climate change: the role of the green infrastructure. Built Environ 33:115–133. https://doi.org/10.2148/benv.33.1.115

Giovannini L, Zardi D, de Franceschi M (2011) Analysis of the urban thermal fingerprint of the city of Trento in the Alps. J Appl Meteorol Climatol 50:1145–1162. https://doi.org/10.1175/2 010JAMC2613.1

GLA (2006) London's urban heat island: a summary for decision makers, London

Gómez-Baggethun E, Barton DN (2013) Classifying and valuing ecosystem services for urban planning. Ecol Econ 86:235–245. https://doi.org/10.1016/j.ecolecon.2012.08.019

Grêt-Regamey A, Celio E, Klein TM, Wissen Hayek U (2013) Understanding ecosystem services trade-offs with interactive procedural modeling for sustainable urban planning. Landsc Urban Plan 109:107–116. https://doi.org/10.1016/j.landurbplan.2012.10.011

Grimsditch G (2011) Ecosystem-based adaptation in the urban environment. In: Otto-Zimmermann K (ed) Resilient cities. Local sustainability, vol 1. Springer, Dordrecht

Guerry AD, Polasky S, Lubchenco J et al (2015) Natural capital and ecosystem services inform-ing decisions: from promise to practice. Proc Natl Acad Sci 112:7348–7355. https://doi.org/10.1073/pnas.1503751112

Haase D, Larondelle N, Andersson E et al (2014) A quantitative review of urban ecosystem ser-vice assessments: concepts, models, and implementation. Ambio 43:413–433. https://doi.org/10.1007/s13280-014-0504-0

Haines-Young R, Potschin M (2010) The links between biodiversity, ecosystem services and human well-being. In: Raffaelli D, Frid C (eds) Ecosystems ecology: a new synthesis. Cambridge University Press, Leiden, pp 110–139

Hansen R, Frantzeskaki N, McPhearson T et al (2015) The uptake of the ecosystem services concept in planning discourses of European and American cities. Ecosyst Serv 12:228–246. https://doi.org/10.1016/j.ecoser.2014.11.013

Harrison PA, Dunford R, Barton DN et al (2017) Selecting methods for ecosystem service assess-ment: a decision tree approach. Ecosyst Serv. https://doi.org/10.1016/j.ecoser.2017.09.016

Hauck J, Görg C, Varjopuro R et al (2013a) Benefits and limitations of the ecosystem services concept in environmental policy and decision making: some stakeholder perspectives. Environ Sci Policy 25:13–21. https://doi.org/10.1016/j.envsci.2012.08.001

Hauck J, Schweppe-Kraft B, Albert C et al (2013b) The promise of the ecosystem services concept for planning and decision-making. GAIA 22:232–236

Heidrich O, Dawson RJ, Reckien D, Walsh CL (2013) Assessment of the climate preparedness of 30 urban areas in the UK. Clim Change 120:771–784. https://doi.org/10.1007/s10584-013-0846-9

Howard E (1902) Garden cities of to-morrow. In: Howard E, Osborn FJ, Munford L (eds) Garden cities of to-morrow. Reprinted ed. with a preface by F.J. Osborn. With an introductory essay by L. Mumford. Faber & Faber, 1951, London

Hübler M, Klepper G, Peterson S (2008) Costs of climate change. Ecol Econ 68:381–393. https://doi.org/10.1016/j.ecolecon.2008.04.010

Jabareen YR (2006) Sustainable urban forms: their typologies, models, and concepts. J Plan Educ Res 26:38–52. https://doi.org/10.1177/0739456X05285119

Jacobs S, Dendoncker N, Martín-López B et al (2016) A new valuation school: integrating diverse values of nature in resource and land use decisions. Ecosyst Serv 22:213–220. https://doi.org/10.1016/j.ecoser.2016.11.007

Jacobson CR (2011) Identification and quantification of the hydrological impacts of impervious-ness in urban catchments: a review. J Environ Manage 92:1438–1448. https://doi.org/10.1016/j.jenvman.2011.01.018

Jopke C, Kreyling J, Maes J, Koellner T (2015) Interactions among ecosystem services across Europe: bagplots and cumulative correlation coefficients reveal synergies, trade-offs, and regional patterns. Ecol Indic 49:46–52. https://doi.org/10.1016/j.ecolind.2014.09.037

Kabisch N (2015) Ecosystem service implementation and governance challenges in urban green space planning – The case of Berlin, Germany. Land use policy 42:557–567. https://doi.org/10.1016/j.landusepol.2014.09.005

Kabisch N, Haase D (2014) Green justice or just green? Provision of urban green spaces in Berlin, Germany. Landsc Urban Plan 122:129–139. https://doi.org/10.1016/j.landurbplan.2013.11.016

Kabisch N, Strohbach MW, Haase D, Kronenberg J (2016) Urban green space availability in European cities. Ecol Indic 70:586–596. https://doi.org/10.1016/j.ecolind.2016.02.029

Kabisch N, van den Bosch M, Lafortezza R (2017) The health benefits of nature-based solutions to urbanization challenges for children and the elderly – A systematic review. Environ Res 159:362–373. https://doi.org/10.1016/j.envres.2017.08.004

Kain J-H, Larondelle N, Haase D, Kaczorowska A (2016) Exploring local consequences of two land-use alternatives for the supply of urban ecosystem services in Stockholm year 2050. Ecol Indic 70:615–629. https://doi.org/10.1016/j.ecolind.2016.02.062

Kazmierczak, A (2012) Heat and social vulnerability in Greater Manchester. In: A Risk-Response Case Study. The University of Manchester, EcoCities

Kenny GP, Yardley J, Brown C et al (2010) Heat stress in older individuals and patients with common chronic diseases. Cmaj 182:1053–1060. https://doi.org/10.1503/cmaj.081050

Koomen E, Diogo V (2015) Assessing potential future urban heat island patterns following climate scenarios, socio-economic developments and spatial planning strategies. Mitig Adapt Strateg Glob Chang. https://doi.org/10.1007/s11027-015-9646-z

Koschke L, Fürst C, Frank S, Makeschin F (2012) A multi-criteria approach for an integrated land-cover-based assessment of ecosystem services provision to support landscape planning. Ecol Indic 21:54–66. https://doi.org/10.1016/j.ecolind.2011.12.010

Kremer P, Hamstead ZA (2016) The value of urban ecosystem services in New York City: a spatially explicit multicriteria analysis of landscape scale valuation scenarios. Environ Sci Policy 62:57–68. https://doi.org/10.1016/j.envsci.2016.04.012

Kremer P, Hamstead ZA, McPhearson T (2013) A social-ecological assessment of vacant lots in New York City. Landsc Urban Plan 120:218–233. https://doi.org/10.1016/j.landurbplan.2013.05.003

Kremer P, Hamstead Z, Haase D et al (2016) Key insights for the future of urban ecosystem services research. Ecol Soc 21:29. https://doi.org/10.5751/ES-08445-210229

Kumar P, Geneletti D (2015) How are climate change concerns addressed by spatial plans? An evaluation framework, and an application to Indian cities. Land use policy 42:210–226. https://doi.org/10.1016/j.landusepol.2014.07.016

Lafortezza R, Davies C, Sanesi G, Konijnendijk C (2013) Green Infrastructure as a tool to support spatial planning in European urban regions. iForest – Biogeosci For 6:102–108. https://doi.org/10.3832/ifor0723-006

Langemeyer J, Gómez-Baggethun E, Haase D et al (2016) Bridging the gap between ecosystem service assessments and land-use planning through Multi-Criteria Decision Analysis (MCDA). Environ Sci Policy 62:45–56. https://doi.org/10.1016/j.envsci.2016.02.013

Larondelle N, Haase D (2013) Urban ecosystem services assessment along a rural-urban gradient: a cross-analysis of European cities. Ecol Indic 29:179–190. https://doi.org/10.1016/j.ecolind.2012.12.022

Lee A, Jordan H, Horsley J (2015) Value of urban green spaces in promoting healthy living and wellbeing: prospects for planning. Risk Manag Healthc Policy:131. https://doi.org/10.2147/RMHP.S61654

Lin BB, Gaston KJ, Fuller RA et al (2017) How green is your garden?: Urban form and socio-demographic factors influence yard vegetation, visitation, and ecosystem service benefits. Landsc Urban Plan 157:239–246. https://doi.org/10.1016/j.landurbplan.2016.07.007

Liquete C, Piroddi C, Macías D et al (2016) Ecosystem services sustainability in the Mediterranean Sea: assessment of status and trends using multiple modelling approaches. Sci Rep 6:34162. https://doi.org/10.1038/srep34162

Liu W, Chen W, Peng C (2014) Assessing the effectiveness of green infrastructures on urban flooding reduction: a community scale study. Ecol Modell 291:6–14. https://doi.org/10.1016/j.ecolmodel.2014.07.012

Luederitz C, Brink E, Gralla F et al (2015) A review of urban ecosystem services: six key challenges for future research. Ecosyst Serv 14:98–112. https://doi.org/10.1016/j.ecoser.2015.05.001

MA (2005) Ecosystems and human well-being: synthesis. A report of the millenium ecosystem assessement. Island Press, Washington, DC

Mace GM (2014) Whose conservation? Science 345:1558–1560. https://doi.org/10.1126/science.1254704

Mace GM (2016) Ecosystem services: where is the discipline heading? In: Potschin M, Haines-Young R, Fish R, Turner RK (eds) Routledge handbook of ecosystem services. Routledge, pp 602–606

Maczka K, Matczak P, Pietrzyk-Kaszyńska A et al (2016) Application of the ecosystem services concept in environmental policy – A systematic empirical analysis of national level policy documents in Poland. Ecol Econ 128:169–176. https://doi.org/10.1016/j.ecolecon.2016.04.023

Maes J, Egoh B, Willemen L et al (2012) Mapping ecosystem services for policy support and decision making in the European Union. Ecosyst Serv 1:31–39. https://doi.org/10.1016/j.ecoser.2012.06.004

Maes J, Liquete C, Teller A et al (2016) An indicator framework for assessing ecosystem services in support of the EU Biodiversity Strategy to 2020. Ecosyst Serv 17:14–23. https://doi.org/10.1016/j.ecoser.2015.10.023

Mathey J, Rößler S, Banse J et al (2015) Brownfields as an element of green infrastructure for implementing ecosystem services into urban areas. J Urban Plan Dev 141:1–13. https://doi.org/10.1061/(ASCE)UP.1943-5444.0000275

McDermott M, Mahanty S, Schreckenberg K (2013) Examining equity: a multidimensional framework for assessing equity in payments for ecosystem services. Environ Sci Policy 33:416–427. https://doi.org/10.1016/j.envsci.2012.10.006

McDonough K, Hutchinson S, Moore T, Hutchinson JMS (2017) Analysis of publication trends in ecosystem services research. Ecosyst Serv 25:82–88. https://doi.org/10.1016/j.ecoser.2017.03.022

Mckenzie E, Posner S, Tillmann P et al (2014) Understanding the use of ecosystem service knowledge in decision making: lessons from international experiences of spatial planning. Environ Plan C Gov Policy 32:320–340. https://doi.org/10.1068/c12292j

McPhearson T, Kremer P, Hamstead ZA (2013) Mapping ecosystem services in New York City: applying a social-ecological approach in urban vacant land. Ecosyst Serv 5:11–26. https://doi.org/10.1016/j.ecoser.2013.06.005

McPhearson T, Hamstead ZA, Kremer P (2014) Urban ecosystem services for resilience planning and management in New York City. Ambio 43:502–515. https://doi.org/10.1007/s13280-014-0509-8

McPhearson T, Andersson E, Elmqvist T, Frantzeskaki N (2015) Resilience of and through urban ecosystem services. Ecosyst Serv 12:152–156. https://doi.org/10.1016/j.ecoser.2014.07.012

McPherson EG, Nowak D, Heisler G et al (1997) Quantifying urban forest structure, function, and value: the Chicago Urban forest climate project. Urban Ecosyst 1:49–61. https://doi.org/10.1023/A:1014350822458

Mullaney J, Lucke T, Trueman SJ (2015) A review of benefits and challenges in growing street trees in paved urban environments. Landsc Urban Plan 134:157–166. https://doi.org/10.1016/j.landurbplan.2014.10.013

Müller N, Kuttler W, Barlag A-B (2014) Counteracting urban climate change: adaptation measures and their effect on thermal comfort. Theor Appl Climatol 115:243–257. https://doi.org/10.1007/s00704-013-0890-4

Munang R, Thiaw I, Alverson K et al (2013) Using ecosystem-based adaptation actions to tackle food insecurity. Environ Sci Policy Sustain Dev 55:29–35. https://doi.org/10.1080/00139157.2013.748395

Nassauer JI, Raskin J (2014) Urban vacancy and land use legacies: a frontier for urban ecological research, design, and planning. Landsc Urban Plan 125:245–253. https://doi.org/10.1016/j.landurbplan.2013.10.008

Naumann S, Anzaldua G, Berry P, et al (2011) Assessment of the potential of ecosystem-based approaches to climate change adaptation and mitigation in Europe. Final report to the European Commission, DG Environment

Niemelä J, Saarela S-R, Söderman T et al (2010) Using the ecosystem services approach for better planning and conservation of urban green spaces: a Finland case study. Biodivers Conserv 19:3225–3243. https://doi.org/10.1007/s10531-010-9888-8

Norton BA, Coutts AM, Livesley SJ et al (2015) Planning for cooler cities: a framework to prioritise green infrastructure to mitigate high temperatures in urban landscapes. Landsc Urban Plan 134:127–138. https://doi.org/10.1016/j.landurbplan.2014.10.018

Oke TR (1988) Street design and urban canopy layer climate. Energy Build 11:103–113. https://doi.org/10.1016/0378-7788(88)90026-6

Olander LP, Johnston RJ, Tallis H et al (2018) Benefit relevant indicators: ecosystem services measures that link ecological and social outcomes. Ecol Indic 85:1262–1272. https://doi.org/10.1016/j.ecolind.2017.12.001

Ortiz MSO, Geneletti D (2018) Assessing mismatches in the provision of Urban ecosystem services to support spatial planning: a case study on recreation and food supply in Havana, Cuba. Sustainability 10:2165. https://doi.org/10.3390/su10072165

Palmer MA, Lettenmaier DP, Poff NL et al (2009) Climate change and river ecosystems: protection and adaptation options. Environ Manage 44:1053–1068. https://doi.org/10.1007/s00267-009-9329-1

Paracchini ML, Zulian G, Kopperoinen L et al (2014) Mapping cultural ecosystem services: a framework to assess the potential for outdoor recreation across the EU. Ecol Indic 45:371–385. https://doi.org/10.1016/j.ecolind.2014.04.018

Pauleit S, Liu L, Ahern J, Kazmierczak A (2011) Multifunctional green infrastructure planning to promote ecological services in the city. In: Urban ecology. Oxford University Press, Oxford, pp 272–285

Persson C (2013) Deliberation or doctrine? Land use and spatial planning for sustainable development in Sweden. Land use policy 34:301–313. https://doi.org/10.1016/j.landusepol.2013.04.007

Piwowarczyk J, Kronenberg J, Dereniowska MA (2013) Marine ecosystem services in urban areas: do the strategic documents of polish coastal municipalities reflect their importance? Landsc Urban Plan 109:85–93. https://doi.org/10.1016/j.landurbplan.2012.10.009

Posner S, McKenzie E, Ricketts TH (2016) Policy impacts of ecosystem services knowledge. Proc Natl Acad Sci 113:1760–1765. https://doi.org/10.1073/pnas.1502452113

Potchter O, Cohen P, Bitan A (2006) Climatic behavior of various urban parks during hot and humid summer in the mediterranean city of Tel Aviv, Israel. Int J Climatol 26:1695–1711. https://doi.org/10.1002/joc.1330

Pueffel C, Haase D, Priess JA (2018) Mapping ecosystem services on brownfields in Leipzig, Germany. Ecosyst Serv 30:73–85. https://doi.org/10.1016/j.ecoser.2018.01.011

Rall EL, Kabisch N, Hansen R (2015) A comparative exploration of uptake and potential application of ecosystem services in urban planning. Ecosyst Serv 16:230–242. https://doi.org/10.1016/j.ecoser.2015.10.005

Raymond CM, Frantzeskaki N, Kabisch N et al (2017) A framework for assessing and implementing the co-benefits of nature-based solutions in urban areas. Environ Sci Policy 77:15–24. https://doi.org/10.1016/j.envsci.2017.07.008

Reckien D, Flacke J, Dawson RJ et al (2014) Climate change response in Europe: what's the reality? Analysis of adaptation and mitigation plans from 200 urban areas in 11 countries. Clim Change 122:331–340. https://doi.org/10.1007/s10584-013-0989-8

Rodríguez JP, Beard TDJ, Bennett EM et al (2006) Trade-offs across space, time, and ecosystem services. Ecol Soc 11:28

Rozas-Vásquez D, Fürst C, Geneletti D, Almendra O (2018) Integration of ecosystem services in strategic environmental assessment across spatial planning scales. Land Use Policy 71:303–310. https://doi.org/10.1016/j.landusepol.2017.12.015

Ruckelshaus M, McKenzie E, Tallis H et al (2015) Notes from the field: lessons learned from using ecosystem service approaches to inform real-world decisions. Ecol Econ 115:11–21. https://doi.org/10.1016/j.ecolecon.2013.07.009

Saarikoski H, Mustajoki J, Barton DN et al (2016) Multi-criteria decision analysis and cost-benefit analysis: comparing alternative frameworks for integrated valuation of ecosystem services. Ecosyst Serv:0–1. https://doi.org/10.1016/j.ecoser.2016.10.014

Sanon S, Hein T, Douven W, Winkler P (2012) Quantifying ecosystem service trade-offs: the case of an urban floodplain in Vienna, Austria. J Environ Manage 111:159–172. https://doi.org/10.1016/j.jenvman.2012.06.008

Santos-Martin F, Viinikka A, Mononen L et al (2018) Creating an operational database for eco-systems services mapping and assessment methods. One Ecosyst 3:e26719. https://doi.org/10.1016/j.jlp.2005.07.001

Säumel I, Weber F, Kowarik I (2016) Toward livable and healthy urban streets: roadside vegetation provides ecosystem services where people live and move. Environ Sci Policy 62:24–33. https://doi.org/10.1016/j.envsci.2015.11.012

Schleyer C, Görg C, Hauck J, Winkler KJ (2015) Opportunities and challenges for mainstreaming the ecosystem services concept in the multi-level policy-making within the EU. Ecosyst Serv 16:174–181. https://doi.org/10.1016/j.ecoser.2015.10.014

Schröter M, Remme RP, Hein L (2012) How and where to map supply and demand of ecosystem services for policy-relevant outcomes? Ecol Indic 23:220–221. https://doi.org/10.1016/j.ecolind.2012.03.025

Schröter M, Albert C, Marques A et al (2016) National ecosystem assessments in Europe: a review. Bioscience 66:813–828. https://doi.org/10.1093/biosci/biw101

Schwarz N, Bauer A, Haase D (2011) Assessing climate impacts of planning policies – An estimation for the urban region of Leipzig (Germany). Environ Impact Assess Rev 31:97–111. https://doi.org/10.1016/j.eiar.2010.02.002

Sen A (2009) The idea of justice. Penguin Books, London

Shashua-Bar L, Hoffman ME (2000) Vegetation as a climatic component in the design of an urban street. Energy Build 31:221–235. https://doi.org/10.1016/S0378-7788(99)00018-3

Skelhorn C, Lindley S, Levermore G (2014) The impact of vegetation types on air and surface temperatures in a temperate city: a fine scale assessment in Manchester, UK. Landsc Urban Plan 121:129–140. https://doi.org/10.1016/j.landurbplan.2013.09.012

Smith P, Ashmore MR, Black HIJ et al (2013) REVIEW: the role of ecosystems and their management in regulating climate, and soil, water and air quality. J Appl Ecol 50:812–829. https://doi.org/10.1111/1365-2664.12016

Souch CA, Souch C (1993) The effect of trees on summertime below canopy urban climates: a case study Bloomington, Indiana. J Arboric 19:303–312

Spake R, Lasseur R, Crouzat E et al (2017) Unpacking ecosystem service bundles: towards predictive mapping of synergies and trade-offs between ecosystem services. Glob Environ Chang 47:37–50. https://doi.org/10.1016/j.gloenvcha.2017.08.004

Stessens P, Khan AZ, Huysmans M, Canters F (2017) Analysing urban green space accessibility and quality: a GIS-based model as spatial decision support for urban ecosystem services in Brussels. Ecosyst Serv 28:328–340. https://doi.org/10.1016/j.ecoser.2017.10.016

Strohbach MW, Haase D (2012) Above-ground carbon storage by urban trees in Leipzig, Germany: analysis of patterns in a European city. Landsc Urban Plan 104:95–104. https://doi.org/10.1016/j.landurbplan.2011.10.001

Syrbe RU, Walz U (2012) Spatial indicators for the assessment of ecosystem services: providing, benefiting and connecting areas and landscape metrics. Ecol Indic 21:80–88. https://doi.org/10.1016/j.ecolind.2012.02.013

Taha H, Akbari H, Rosenfeld A (1991) Heat island and oasis effects of vegetative canopies: micro-meteorological field-measurements. Theor Appl Climatol 44:123–138. https://doi.org/10.1007/BF00867999

Tallis H, Polasky S (2009) Mapping and valuing ecosystem services as an approach for conservation and natural-resource management. Ann NY Acad Sci 1162:265–283. https://doi.org/10.1111/j.1749-6632.2009.04152.x

Tang Z, Brody SD, Quinn C et al (2010) Moving from agenda to action: evaluating local climate change action plans. J Environ Plan Manag 53:41–62. https://doi.org/10.1080/09640560903399772

TEEB (2010a) The economics of ecosystems and biodiversity: mainstreaming the economics of nature: a synthesis of the approach, conclusions and recommendations of TEEB

TEEB (2010b) The economics of ecosystem and biodiversity for local and regional policy makers

Tzoulas K, Korpela K, Venn S et al (2007) Promoting ecosystem and human health in urban areas using green infrastructure: a literature review. Landsc Urban Plan 81:167–178. https://doi.org/10.1016/j.landurbplan.2007.02.001

UN-Habitat (2016) Urbanization and development: emerging futures. World Cities Report 2016

Vallecillo S, La Notte A, Polce C et al (2018) Ecosystem services accounting: part I – Outdoor recreation and crop pollination. Publications Office of the European Union, Luxembourg

Vignola R, Locatelli B, Martinez C, Imbach P (2009) Ecosystem-based adaptation to climate change: what role for policy-makers, society and scientists? Mitig Adapt Strateg Glob Chang 14:691–696. https://doi.org/10.1007/s11027-009-9193-6

von Haaren C, Albert C (2011) Integrating ecosystem services and environmental planning: limitations and synergies. Int J Biodivers Sci Ecosyst Serv Manag 7:150–167. https://doi.org/10.1080/21513732.2011.616534

Wamsler C, Luederitz C, Brink E (2014) Local levers for change: mainstreaming ecosystem-based adaptation into municipal planning to foster sustainability transitions. Glob Environ Chang 29:189–201. https://doi.org/10.1016/j.gloenvcha.2014.09.008

Watkins R (2002) The impact of the urban environment on the energy used for cooling buildings. Brunel University, Uxbridge

Wilkinson C, Saarne T, Peterson GD, Colding J (2013) Strategic spatial planning and the ecosystem services concept – An historical exploration. Ecol Soc 18:37. https://doi.org/10.5751/ES-05368-180137

Wolff S, Schulp CJE, Verburg PH (2015) Mapping ecosystem services demand: a review of current research and future perspectives. Ecol Indic 55:159–171. https://doi.org/10.1016/j.ecolind.2015.03.016

Woodruff SC, BenDor TK (2016) Ecosystem services in urban planning: comparative paradigms and guidelines for high quality plans. Landsc Urban Plan 152:90–100. https://doi.org/10.1016/j.landurbplan.2016.04.003

Yu C, Hien WN (2006) Thermal benefits of city parks. Energy Build 38:105–120. https://doi.org/10.1016/j.enbuild.2005.04.003

Zardo L, Geneletti D, Pérez-Soba M, Van Eupen M (2017) Estimating the cooling capacity of green infrastructures to support urban planning. Ecosyst Serv 26:225–235. https://doi.org/10.1016/j.ecoser.2017.06.016

Zimmerman R, Faris C (2011) Climate change mitigation and adaptation in North American cities. Curr Opin Environ Sustain 3:181–187. https://doi.org/10.1016/j.cosust.2010.12.004

Zulian G, Paracchini ML, Maes J, Liquete Garcia MDC (2013) ESTIMAP: ecosystem services mapping at European scale. Publications Office of the European Union, Luxembourg

Zulian G, Stange E, Woods H et al (2018) Practical application of spatial ecosystem service models to aid decision support. Ecosyst Serv 29:465–480. https://doi.org/10.1016/j.ecoser.2017.11.005

Index